£3.25

STEAM YACHTS

STEAM YACHTS

DAVID COULING

B.T. BATSFORD LTD, LONDON

For my grandmother – Martina

Times remembered, long, long ago

First published 1980
© David Couling 1980

All rights reserved. No part of this publication
may be reproduced in any form or by any means
without permission from the publishers.

ISBN 0 7134 1884 2

Filmset in Monophoto Century Schoolbook by
Servis Filmsetting Ltd, Manchester

Printed in Great Britain by
The Anchor Press Ltd, Tiptree, Essex
for the publishers B.T. Batsford Ltd,
4 Fitzhardinge Street, London W1H 0AH

Acknowledgments

I am most grateful to the following for their help and kindness in providing me with information and photographs to include in this book.

Mr Kirk of Cyprus; Mr L.J. Mitchell BA, FLA, Director of Cultural Services, Newport, Isle of Wight; Mr Brian Hillsdon, Historian to the Steam Boat Association of Great Britain; Mr Arthur G. Credland, Keeper, Town Docks Museum, Kingston Upon Hull; Mr E. Sharp of Guernsey; Mr J.L. Rodger, Research Assistant, University of Glasgow; the staff of the Cowes Maritime Museum, Cowes, Isle of Wight; Mr B. Cox; Lady Comperston; Mr G. Davies; Mrs J. Couling and Maldwin Drummond DL, JP.

I should also like to thank Mr William F. Waller, Editorial Director, B.T. Batsford Ltd, for his kindness and understanding.

Author's Note

Every care has been taken to record the details given in this book as accurately as possible, but with so little information at hand, and ships' records differing from one source to another, it has become a difficult task. I would therefore appreciate any further facts or photographs relating to steam yachts, in order to build up a comprehensive file on this subject. Such information may be sent to David Couling, c/o B.T. Batsford Ltd, 4 Fitzhardinge Street, London W1H 0AH.

David Couling, 1980

Introduction

In the winter of 1976, whilst researching for material on a book which I was writing on the Isle of Wight, I was asked by the Director of Cultural Services for the Isle of Wight, to examine a quantity of glass negatives which his department had recently taken charge of. On showing me into a storeroom, I was confronted by twelve large cardboard boxes, packed tight with glass negatives. To me, the excitement of looking at these negatives which had not seen the light of day for the last seventy years, was equal to an archaeologist finding a hoard of Roman coins. I found that three-quarters of the collection were of 12 × 10-inch glass negatives, whilst the remainder were 5 × 4 inches. In all, the collection numbered a little over 1000 plates. Apart from a small number of negatives taken at Osborne Naval College, the remainder related to yachting in and around Cowes, Isle of Wight.

It soon became apparent that those old cardboard boxes contained an immensely important photographic collection, covering the period from 1880, until the outbreak of the Second World War; many would say, the Golden Age of yachting. Yet this beautiful collection contained even more delights, for within the main collection there was a smaller and uniquely interesting set of plates depicting steam yachts. These covered the period from 1880 to 1914, the halycon days of yachting. The negatives were in a sad state of preservation, having been stored either in a cellar or builder's yard. Most were saturated with brick dust and soil, which made necessary the restoration of the plates straight away. To bring the negatives back to their original state took a little over a year.

The collection was the result of many years' work by William Kirk of Cowes. Kirk started as a photographer in 1879, specialising in portrait, marine and landscape subjects. As his business developed, his main theme became marine photography, although other work was carried out. He was patronised by Queen Victoria, her family and many of the Crown Heads of Europe. In all probability, Kirk took his marine photographs in the hope of selling prints, displaying them in his studio window. However, over the years his reputation grew, and commissions to photograph yachts and steam yachts came in thick and fast.

By the early 1890s, Kirk had become one of the foremost marine photographers in the Isle of Wight.

Competition from such giants in marine photography as Frank Beken of Cowes, Debenham of Cowes and G. West & Son of Southsea and Gosport became fiercer. By 1904, Kirk's sons had joined the family firm and continued taking marine photographs after their father's death, until 1939, when the firm appears to have closed down.

The era of the steam yacht lasted less than seventy years, 1843–1918, but within this short space of time some of the most graceful and luxurious steam yachts to be built graced our waters. The yachts were a visible sign of wealth. For the cost of owning a steam yacht was indeed great. The first private steam yacht to be built was the *Menai*, designed by Robert Napier. Built in 1830, she was 120 feet long with a beam of 20 feet and a tonnage of 400. Her engines were the double-side-lever type driving her paddle-wheels, which gave her a nominal speed of 110h.p.

Her owner, Asheton-Smith, paid £20,000 for her, and by today's standards *Menai* would have cost in the order of a million pounds to build. In all Asheton-Smith owned nine steam yachts, including the renowned *Fire King*, the fastest in his fleet. In 1839 he issued an open challenge, that *Fire King* could outrun any steamer afloat, and laid a wager of £5000 to back it up.

Asheton-Smith has one other claim to fame in the annals of steam yachts, in that he resigned from the Royal Yacht Club at Cowes. In a resolution passed in May of 1827, the Club stated that no member should own a steam yacht; if so, he was to be disqualified from membership. This so infuriated Asheton-Smith that he at once resigned from the Royal Yacht Club and commissioned the *Menai* to be built forthwith. It was not until 1843 that the then Royal Yacht Squadron capitulated to steam, and allowed members to own steam yachts of over 100 tons. By the middle of the 1850s the graceful steam yacht had come into her own, taking her place alongside the proved and tried schooners and cutters.

Between 1850 and the outbreak of the First World War shipyards were busily building steam yachts to cater for a growing demand from the wealthy. From this demand came such magnificent yachts as *Valhalla* (1218 tons), *Sunbeam* (334 tons) and *Sapphire* (1207 tons). *Sapphire* was built by John Brown, for the Duke of Bedford in 1912, and can be justly classed as one of the last great thoroughbred steam yachts.

As a young man in 1912, Eric Sharp was asked to attend an interview for the post of Wireless Officer on board the *Sapphire*. Having been interviewed, Eric Sharp was offered the post at the then staggering salary of £3 per week, which was greatly in excess of the sum he had been paid whilst working on the great liners. Apart from his salary, he was instructed to visit leading outfitters in London to be measured for uniforms, at the owners' expense. These consisted of one blue serge suit, one cloth suit, a magnificent overcoat, two yachting caps and two pairs of brown shoes – all of the highest quality. In addition to salary and yacht kit, Eric Sharp was supplied with one pound of tobacco and plenty of matches to light his pipe. As far as Wireless Officer Sharp was concerned, he had reached heaven. The yacht's uniform was to be worn at all times, both on board the yacht and while on shore duty. Under no pretext whatever were black shoes to be worn when on board the yacht, and Eric Sharp was later to find out why.

After arriving at Southampton, Wireless Officer Sharp found the *Sapphire* swinging at her moorings on the gympe, between the docks and Hythe. A motor launch from the pier conveyed him to the *Sapphire*, where he recalls that

> ...I was soon climbing the accommodation ladder with its neatly canvassed ropes. I quickly realised why black footwear was forbidden. The decks were spotless and indeed the whole ship from waterline to masthead shone with cleanliness. I reported to the Master – Captain Elliot – and learned that he and the majority of the Ship's Company had been with the owners for many years, serving on the old *Sapphire* that had been replaced by the new and much larger one. Indeed, only the Wireless Operator, Second Officer and Fourth Engineer were newcomers. The total crew was about 49 and the fact that this number stayed year after year indicated a happy ship. This proved to be the case.
>
> All the deck houses gleamed with polished varnish and my wireless cabin at the after end of the boat deck was in a similar condition. It was a roomy cabin housing the standard Marconi equipment of the era, i.e. fixed-gap transmitter, magnetic detector and multiple tuner receiver with a 10-inch induction coil and a 24-volt battery as emergency. The aerial was a standard silicone bronze twin-wire on spreaders and the height of the masts assured me that I could look forward to good ranges. It was up to me to keep all the gear spick and span and in keeping with the rest of the ship.
>
> All the crew lived forward, the officers having their own mess-room and steward. Captain, mates, engineers, wireless operator, carpenter, boatswain and chef constituted the mess. The officers' cabins were in keeping with the ship. I shared a two-berth cabin with the electrical engineer, which was well fitted with a writing desk, cupboards, drawers and individual reading lamps. We were not fed by the owners, but each officer paid the steward 10/- a week and with the proceeds he supplied the food. As very large refrigerators were fitted on board there was no problem in getting food in reasonable quantities. The steward did the shopping in ports frequented by yachts, and it was a common sight on shore to see numbers of stewards easily identified by the typical woven shopping basket. . . .
>
> In due course, orders came through to prepare for sea, and I made up my mind that in spite of the novel situation, I would carry on my wireless duties in the way I had been accustomed to. As luck would have it, I stepped off on the right foot, as the day

before sailing I copied the long coded weather report transmitted by the Eiffel Tower in Paris, whose signal was very strong and carried up to about 1500 miles. I decoded the relevant portions and took them up to the Captain. He wanted to know what it was, and when I explained he said 'Why did I never hear of this before. Her Grace will be delighted.' Needless to say, he received the weather report throughout the commission. But the most important part of my duties was to get the news from Poldhu every night. This meant that I had to be on watch at 2 a.m. and stay there until I had got the lot, which always meant waiting up for the repeat transmission as fading was very bad, especially on the first transmission, But the most important thing was the receipt of the nightly message from the Duke to his wife. He never came near the ship, and I believe he only saw her once while she was on the stocks, but he never failed in his nightly message. . . .

On the run up the Irish Sea, I naturally enquired as to the likely composition of the party we were to pick up, and learnt to my astonishment that there was no party, but only the Duchess, her personal companion/maid and her Pekingese dog, and that the latter ruled the ship with a rod of iron! It transpired that Her Grace was one of the greatest authorities on bird life, especially that connected with the sea, and that the yacht was wanted to ferry her from one bird sanctuary to another and from one remote bird-haunted rock or sea-loch to another. So it came about that the whole commission was spent amongst the wild and beautiful coasts of Scotland, the Hebrides, Orkneys and Shetlands, with the exposed and wind-swept Fair Isle in the entrance of the Pentland Firth as a focal point. Here the Duchess had a cottage. She loved the tough Scandinavian-looking inhabitants, who wrested a living from the hard soil and the seas around them. When we were in Southampton, I had noticed a number of packing cases and assorted bundles coming aboard to be stowed in the hold forward. At Fair Isle, their purpose was revealed – they contained comforts of all kinds for distribution among the inhabitants. It was Her Grace's invariable custom to do this, and the people were everlasting grateful to her for the joy and comfort she brought. Having landed the Duchess and her gifts, and seen them safely on their way up the cliff path the *Sapphire* proceeded to Lerwick for the night. Here I copied the news bulletin and the Duke's message and next day we returned to the island and delivered them. This routine was kept up as long as she was on the island. Lerwick at this time was a very interesting place as the herring shoals had moved North and with them came the drifters and the fisher lassies who cleaned and gutted the fish and prepared them for their conversion to succulent kippers. . . .

From Lerwick we might go anywhere where birds were to be found, to be studied, photographed and painted. Under no circumstances was a bird shot, but painstakingly stalked and studied through field glasses. I well remember the day we steamed up Loch Eribol on the North Coast, where the Duchess found a nesting pair of Golden Eagles. Clad in old tweed skirts with a cap on her head held in place by a couple of hatpins, she lay in the heather for days studying the wonderful sight. We seldom went into busy ports – Oban and Lerwick were about the limit, but Tobermory was very attractive and we anchored over the spot where the famous Tobermory galleon is supposed to rest. But my memories of the cruises are of the incredible beauty of the Scottish coasts from the majesty of the mighty cliffs of Foula, the serenity of the Isles of the Sea, the pure, golden sands of the smaller uninhabited islands of the Outer Hebrides where we were able to gather wonderful cockles, while Her Grace painted her pictures of sea birds in their undisturbed surroundings. As vivid as all this is the remembrance of the terrible seas of the Pentland Firth, where all ships had to fight for their lives. . . .

Most of the crew had served on yachts all their lives and it can be imagined the yarns that were spun round the mess table, or in our recreation room over the games of cards. Stories, amusing and scandalous, but always interesting, covering as they did all ranks of Society, from crowned heads downwards. They would have made a fortune for a gossip columnist, but they never went beyond the steel walls of the ship. All hands were fiercely loyal to their owners and it was interesting to note that the appelation 'the owner' was never regarded as anything but right and proper. The most popular owner seemed to be Sir Thomas Lipton, whose stately *Erin* was regarded as a home from home. Everybody was sad at the repeated failure of his *Shamrock* racers to capture the America's Cup.

Of the people who chartered the old *Sapphire* by far the most popular was the late Herr Krupp, the German armaments king. He was a father to all hands, took a great interest in their lives and on mail-day sat down with them on a hatch and discussed their joys and sorrows. Another popular character was a well-known English aristocrat, who on a West Indian cruise insisted that all hands that could be spared should have shore leave when anchored off a town. As the men went down the accommodation ladder each was presented with a golden sovereign and wished a happy evening.

Eric Sharp was indeed a fortunate man to have served upon the classic steam yacht of that time, for by the close of 1912, their twilight was fast approaching. The storm clouds of the First World War were beginning to envelop Europe and by early 1914 private steam yachts were being requisitioned for use in the Royal Navy. Large steam yachts were being used as accommodation ships or hospital ships, whilst smaller craft were used as armed patrol vessels. Enemy action took its toll and by 1919 the age of the beautiful steam yacht had all but come to an end. It had become almost out of the reach of even the very wealthy to run and maintain their yachts. Building costs had sored to such an extent that fewer and fewer were being planned or built. Many owners tried to maintain their yachts, but fuel costs and wages had also risen unacceptably, the result being that many yachts ended their days along river banks, having been converted into houseboats. Although the age of the steam yacht lasted for a mere seventy years, it produced the best in craftsmanship from our shipyards, and this was reflected in the demand from buyers from all parts of the world.

Today, few steam yachts survive, but one or two may be found in the Mediterranean. Even fewer are built. Most notable of recently built ones is the royal yacht *Britannia*, and Aristotle Onassis' yacht *Christina*.

Britannia was constructed in the John Brown shipyards in 1954, to replace the ageing *Victoria & Albert III*. Built of steel, she has a gross tonnage of 5769 and is powered by two water-tube oil boilers, her engine being two geared turbines. She has a length overall of 412 feet 3 inches, a beam of 55 feet, and a draft of 15 feet 6 inches, and can attain a speed of 22 knots. In the event of war, *Britannia* can be converted into a hospital ship within a very short period of time.

Christina's hull was constructed in 1943 as a frigate, her original name being HMS *Stormont*. Becoming surplus after the war, Onassis converted her into a luxurious steam yacht. She has a gross tonnage of 1602, a length of 325 feet 3 inches, with a beam of 36 feet 5 inches, and a draft of 15 feet 2 inches. Powered by two water-tube oil boilers, she can reach a speed of 21 knots.

The era of the majestic steam yacht has passed, but fortunately, through the lens of Kirk of Cowes and other great marine photographers, we can recall that fine age. Money was then no object to pay for the best materials and the finest craftsmen to build those elegant ladies of the sea.

Shipwrights, the mainstay of any yacht builders, taking a morning break from work in the 1890s, at a shipyard in East Cowes.

Apart from farming, shipbuilding was the main industry on the Isle of Wight for many years.

VICTORIA & ALBERT II 1855

DESIGNER, O. Lang BUILT 1855 by Pembroke Dockyard
LOA 300 feet BEAM 40 feet 3 inches
DRAUGHT 16 feet 6 inches GROSS TONS 2470
MATERIAL wood RIG paddle schooner
SPEED 15.4 knots

Laid down as the *Windsor Castle*, but her name was changed when she was launched in 1855.

Although Sir Oliver Lang designed the royal yacht, Prince Albert took responsibility for the decoration of the state apartments. Painted in white with chintz wallpaper, the apartments were to provide a home-from-home atmosphere away from formal court life ashore. The yacht became a firm favourite with the royal couple and their children, and was frequently used to convey the Queen and Prince Albert to the remoter parts of the British Isles. With the untimely death of the Prince in 1861, the Queen became a recluse, the *Victoria & Albert* being rarely used and moored in home waters. However, in 1863 the royal yacht made a welcome trip abroad to bring Princess Alexandra to England for her forthcoming marriage to the then Prince of Wales, later to become King Edward VII.

By the early 1870s the Queen had once again renewed her interest in both the sea and her beloved yacht, which she continued to use until her death in 1901. The Queen's last voyage on the royal yacht was on a state visit to Ireland in 1900, but by the end of her tour it was seen that the Queen was becoming very weak. By the end of the year, Queen Victoria had returned to her retreat at Osborne House, Isle of Wight, gravely ill. On 22 January 1901, Queen Victoria died at Osborne and so too had the *Victoria & Albert II* come to her end, after giving service spanning sixty-six years.

On 1 February 1901 the Queen's coffin was taken on

board the royal yacht *Alberta* in company with the *Victoria & Albert II*. The Lords of the Admiralty steam yacht *Enchantress*, the German royal yacht *Hohenzollern*, and the Queen's body left Cowes en route for Portsmouth, accompanied by eight destroyers.

At Portsmouth, the Queen's coffin was taken on to Windsor and thence to Frogmore for burial. By the end of February the *Victoria & Albert II* had been laid up and in 1904 she was taken to the breakers' yard to be scrapped. Those last survivals of the Victorian era had sadly disappeared.

Above The steam launch *Isis* built by Yarrow's men in their spare time during the late 1860s. This proved to be the turning point for Yarrow who, after building *Isis*, went into the production of small steam launches.

The royal yacht *Alberta*, tender to the royal yacht *Victoria & Albert II*, was built at Pembroke in 1863 to replace the royal tender *Fairy*. She had a length of 160 feet with a beam of 22 feet 8 inches and a displacement of 370 tons. The *Alberta* had the sad duty of carrying the remains of Queen Victoria to the mainland from the Isle of Wight, in January 1901.

Right One of the many steam launches built by Yarrow in the 1870s. This type was used extensively on inland waterways.

STELLA c.1870

The *Stella*, a 45-foot twin-screw launch with a draught of $2\frac{1}{2}$ feet, was built by the then young firm of Yarrow and Hedley, on the Isle of Dogs, London. The cost of the *Stella* would have been approximately £400, but in the formative years Yarrow often worked at a loss. In Lady Yarrow's book, *Alfred Yarrow, His Life and Work*, she recalls the early financial difficulties of the firm. A contemporary advertisement read:

> STEAM LAUNCHES – Anyone wanting a steam launch would be well served if they come to Yarrow and Hedley, Isle of Dogs.

Lady Yarrow went on:

> This was the last resource. Without capital, and with the business considerably in debt, but with increased experience, Yarrow awaited developments. Within three days of the appearance of this advertisement, an old gentleman, Colonel Halpin, came to the yard and placed an order for a steam launch, 24ft in length, to be fitted with a small cabin to hold four people. It was to have a single-cylinder engine and a vertical boiler of 3 horse power. The price was to be £145. This little boat was built in three months and cost £200 so that the financial outlook had not yet improved.

Later in the 1890s, Yarrow was to move his works to Clydeside and there became one of Britain's biggest ship builders.

Above right Possibly a more expensive Yarrow steam launch, incorporating a canopy over the boiler. As in the following illustration the launch was built to be used on the river.

Right An early Yarrow steam launch, built for river use in 1870. Note the exposed boiler.

A 32-foot steam launch constructed by J. Samuel White's shipyard, for the royal yacht *Victoria & Albert II*. This was one of several built by the firm for the royal yacht over a number of years.

Two early wooden steam torpedo boats, built by J. Samuel White of Cowes, c.1870s. In some cases steam yachts were built on the torpedo boat design, in order to gain extra speed. A notable example is that of the steam yacht *Scud*.

MARGARITA 1873

BUILT 1873 by P.B. Seth & Co., Rutherglen, Glasgow
LOA 187 feet   MATERIAL iron
ENGINE compound 22, 44 × 27 by A. Camble & Co., Glasgow
GROSS TONS 489

Although *Margarita* was built in 1873, she was not purchased by her first owner, Mr H. Bishaffshein, until 1875, her name being changed to *Cavalier* in 1874. Henry Bishaffshein kept *Cavalier* for less than a year when she was purchased by Mr N.R. Shaw-Stuart. Mr Shaw-Stuart kept her in service until 1903 when she was bought by Sir S.H. Lever. She remained in British waters under the ownership of Sir S.H. Lever until 1910 when she was sold to the United States of America.

Right A Yarrow steam launch, ready for delivery at their shipyard at the Isle of Dogs c.1875.

BLUNDERBUS 1874

BUILT 1874 by J.G. Fay & Co., Northam, Southampton
LOA 45 feet   BEAM 10 feet 4 inches   MATERIAL wood
ENGINE twin cylinder 9, 9 × 9, later converted to compound 9, $14\frac{1}{2} \times 9$
RIG twin screw yawl

Blunderbus, an early example of a steam yacht, was built by Fay & Co. who were mainly involved in laying up yachts, although at one time they did take over another yacht-building firm, and they also built steam tugs for a short time. *Blunderbus*, c.1875, is a little unusual as she was constructed as a yawl. She is seen blowing off steam, and to the right is a Thames sailing barge. Her owners included Count Batthyany, 1875–81; Duke of Bedford, 1885–92; Surgeon Lt-Col. G. Ryan, 1894–6; I. Reed, 1897–1900; E.W. Balne, 1901; and C. Van Raalte, 1902–19.

SUNRISE 1877

BUILT 1877 by J. Samuel White, Cowes
LOA 133 feet 8 inches   BEAM 24 feet 3 inches
DRAUGHT 13 feet 3 inches
ENGINE compound 19, 33 × 18 by John Perm, Greenwich
MATERIAL composite   GROSS TONS 209

Sunrise was not a pure steam yacht, but a three-masted square rigger with a steam-powered engine for auxiliary use only. She was constructed for Lord Ashburton who owned her until 1896, when she was sold to E.J. Coope, who belonged to the family of brewers. She had had new boilers fitted in 1897. She stayed with the Coope family until 1904, when she was sold once again. Her new owner, C.H.L. Cazelet, kept her only for a year when she was sold to Mr A. Reynall-Pack. *Sunrise* stayed with Reynall-Pack for five years, and in 1912 she was purchased for commercial interest in France. By 1913 she had been renamed *Yves de Kerguelen*.

The illustration shows *Sunrise* whilst in the ownership of C.H.L. Cazelet.

AMY 1877

BUILT Barrow LENGTH 161 feet RIG screw schooner
MATERIAL iron GROSS TONS 287
DESIGNED St Clair Byrne
ENGINE original engines not known. In 1896 new engines fitted by J. Jones of Liverpool. Triple expansion 13, 22, 25 × 24

Details of this vessel seem scarce. She was designed by St Clair Byrne, one of the foremost yacht designers of the mid-nineteenth century, whose most outstanding achievement was that of Earl Brassey's auxiliary three-masted top-sail schooner, *Sunbeam*.

Amy's name was later changed to *Jason* whilst under the ownership of Mr Frank Bibby, but by the beginning of the First World War, when she was hired by the Admiralty, her name had reverted to *Amy*. Armed with a 12-pounder and a 6-pounder, *Amy* was used in British waters as an auxiliary patrol vessel until March 1919, when she was returned to her owner.

ORIENT 1878

BUILT 1878 by J. Reid & Co., Port Glasgow
LOA 100 feet   BEAM 13 feet   MATERIAL iron
ENGINE compound 12, 24 × 21 by Walker Henderson

Orient at anchor off Cowes *c*.1908, showing the classic design of steam yachts of the 1870s.

It was always considered a good berth to crew on board a private steam yacht and the crew would proudly sport their blue jerseys with the yacht's name in white across their chests. Rivalry was always high between different ships' crews to maintain the best turned-out craft.

Orient's owners included P. Mackesson, 1880–1, and Sir William Mackesson, 1885–94, of Mackesson stout fame; A. Hitchman, 1895–6; O. Pederson, 1897–1906. She was then sold to Greek commercial interests in 1907, and her name was changed to *Salamina*.

The Liquid Fuel Engineering Company's subsidiary works at Poole, Dorset. Several of their steam launches can be seen moored beside their wharf.

Above An early photograph of a LIFU steam launch c.1880, on the river Medina at East Cowes. In the background can be seen Souter Shipyard, an old-established island firm.

GLADYS 1883

BUILT 1883 by J. Reid & Co., Glasgow
LOA 128 feet 15 inches   BEAM 17 feet 5 inches
DRAUGHT 9 feet 1 inch   MATERIAL iron
RIG screw schooner   GROSS TONS 151.25
ENGINE compound 14, 28 × 21

The general design of the *Gladys* was typical of her day, being a neat compact yacht giving good service during her time at sea. In her career she had over a dozen owners, but perhaps one of the most touching moments in the yacht's life occurred on Wednesday 10 May 1889. Berthed in the river Medina, her owner, Mr C.M. Burt, had dressed the *Gladys* overall with bunting fluttering from the mast and stays. The occasion was his daughter's wedding day. *Gladys* stayed in service until the early 1950s, then in 1952 she was converted to a houseboat and records cease.

GLADIATOR 1886

BUILT 1886 by Ramage & Ferguson Ltd, of Leith
LOA 119 feet 6 inches   BEAM 20 feet 1 inch
DRAUGHT 11 feet 7 inches   MATERIAL iron
ENGINE triple expansion $9\frac{1}{4}$, 15, $24\frac{1}{2}$ × 18, by Ramage & Ferguson
GROSS TONS 164

Gladiator was built for Mr R. Martin, who owned her only for one year before she was sold to Sir J.W. Kelk of Marlborough. The Steam Pilot Boat Co. of Cardiff finally purchased her in 1913, but only after she had been in service with six other owners – C.L. Orr-Ewing, 1895–6; Major Orr-Ewing, 1897–9; R.G. Calvert of Bordeaux, 1900; Marquis de Polignac, 1901; Dr J.B. Charcot, Le Havre, 1902; H.W.S. Gray of Southampton, 1904–12.

Whenever *Gladiator* changed hands, her name was also changed. More often than not a ship's name was usually retained by the new owner. Sadly, in 1914 she was lost at sea whilst on pilot duty.

MORVEN 1886

BUILT 1886 by Caird & Co., Greenock
LOA 157 feet 5 inches   BEAM 22 feet 1 inch
ENGINE triple expansion 14, 22, 36 × 30,
by Caird & Co., Greenock

Morven had a comparatively short career under British ownership. Built for J. Rolls-Hoare, who owned her from 1888–98, she was then sold to the Earl of Latham. The Earl kept *Morven* until 1902, when she was sold to Mr C.H. Potter. By 1909 she had once again been sold, this time to a French owner, her name having been changed to *Henriette*.

The illustration above shows the *Morven* at anchor off Cowes in 1897.

DOROTHEA 1887

BUILT 1887 by Oswald Mardaunt & Co., Woolston, Southampton
LOA 56 feet 3 inches   BEAM 11 feet 7 inches
MATERIAL composite
ENGINE compound 7, 14 × 12, by Oswald Mardaunt
GROSS TONS 28

This photograph of 24 January 1911, unlike many taken by Kirk of Cowes, was commissioned by the owner, Comte M. de Grand S. d'Hauterises, who can be seen standing on the foredeck with the crew. He was the last known owner, as records of her cease after 1914.

She previously had five owners: Major E.F. Coates of Southampton, 1886–90; A. Spooner, 1899–1902 (there appears to be no record of ownership during the intervening years); C.T. Wilkinson, 1904–7; T.A. Dicker, 1908–9; G. Marvin, 1910–11.

Whilst under the ownership of the Comte, she was renamed *Hélène*.

WILDFIRE 1887

BUILT 1887 by J. Samuel White
MATERIAL steel   LOA 91 feet   ENGINE compound

Wildfire was owned by Lt-Gen. C. Bearing until 1891, after which time she had a succession of owners. Notable amongst them was E. Mercedes, who owned her from 1908–9. She was re-named *Mercedes* in 1908, and *Le Tigre* in 1910. *Wildfire* is seen here off Cowes in 1887.

Top Valhalla's massive steering gear, after being reconditioned, having been replaced in its original position.

Above The steering gear of the *Valhalla* being reconditioned by J.S. White of Cowes.

VALHALLA 1892

BUILT 1892 by Ramage & Ferguson   LOA 245 feet
BEAM 37 feet 3 inches   DRAUGHT 20 feet
GROSS TONS 1218   MATERIAL steel
ENGINE T.E. $18\frac{1}{2} \times 27\frac{3}{4} \times 47.33$

The *Valhalla* was probably the only British full-rigged steam yacht built. She is remembered as being one of three British competitors in the Kaiser's Transatlantic Race in 1905. Her companions were *Sunbeam* and *Apache*. *Valhalla* came third overall in a time of 14 days 2 hours, beaten by *Atlantic* and *Hamburg*.

The great yacht made many trips abroad with her third owner, the Earl of Crawford, to catch rare birds. In the First World War she was used by the Royal Navy as a floating workshop. In 1920 she was sold to commercial interests in Spain and became a fruit carrier. She was wrecked and sunk off Cape St Vincent in 1922. *Valhalla* is seen here full-rigged in 1904.

MOONBEAM 1892

BUILT 1892 by W. White & Sons, Cowes, Isle of Wight
LOA 111 feet   BEAM 15 feet 4 inches   GROSS TONS 899
ENGINE triple expansion $8\frac{3}{4}$, $13\frac{3}{4}$, 21×16,
by Wisson & Co., Gloucester

A charming example of an official photograph taken at Cowes Week in 1900, showing *Moonbeam*'s owners with their friends and the ship's company posed for the photographer, Kirk of Cowes. In the background, left, with her bows just jutting out beyond *Moonbeam*'s, is the royal yacht *Victoria & Albert II*, and in the right-hand background can be seen the royal yacht *Victoria & Albert III*.

Originally built for A.R. & J.M. Sladen, the *Moonbeam* was sold to a Mr Leghorn in 1895. Her other names were *Ensa* 1895, *Moonbeam* 1904, *Beq-Hir* 1909, and *Elena* 1922. After having had many owners, she ended her days in France, where records cease in 1950.

Right The Royal Yacht Squadron, Cowes, Isle of Wight, c.1890. Founded in 1815.

In 1827, Asheton-Smith, a member of the Royal Yacht Squadron, resigned in protest over the disapproval of steam-powered craft. It was Asheton-Smith who in 1830 commissioned the first private steam yacht in England, from the designer Robert Napier. Called the *Menai*, she had a length of 120 feet with a beam of 20 feet and a tonnage of 230. She was propelled by paddle and had a nominal h.p. of 110.

In 1843, the Royal Yacht Squadron capitulated, allowing steam yachts of 100 tons and over.

PALADIN 1895

BUILT 1895 by Forrest & Son, of Wyvenhoe
LOA 88 feet   MATERIAL galvanised steel
ENGINE compound 7 × 13 inches diameter,
by W. Sissior & Sons, Gloucester
BOILER by Davy Paxton, 120 pounds p.s.i.

Although *Paladin* was quite small compared with other steam yachts of her time, she sports the classic lines of her larger counterparts. In the photograph she displays all the grace and charm of the Victorian age. The proud little steam yacht, underweigh at Cowes, has her owner R. Vogain in a striking pose in the bows.

Paladin was owned by R. Vogain until 1902, and she was sold to a Belgian in 1903. Her name was changed in 1925 to *Tchantches*. As to her ultimate fate, it is not recorded.

The steam yacht *Xarifa* berthed at Cowes in 1894. She is flying the American flag. Her owner at this time was F.M. Singer of New York.

The auxiliary steam yacht *Xarifa* under full sail in 1894.

IVY   1895

BUILT 1895 by Earl Ship Building Co., Hull
LOA 220 feet   GROSS TONS 869   MATERIAL steel
ENGINE twin screw triple expansion

Although *Ivy* was built in England, she spent little time in home waters. Her first owners were Nigerian Coast Protectorate, who kept her until 1905. It is possible that, while in their service, she was used to combat smuggling.

By 1909, *Ivy* was owned by the Government of Southern Nigeria and stayed in their service until 1923, when she was bought by the Sheik of Mahaanerah.

In 1936, *Ivy* came back to British waters to be reconditioned by J. Samuel White of Cowes. After 1936, she does not appear in Lloyd's Register.

The above photograph was taken after she had been reconditioned. It is interesting to note that she has square port holes, instead of the standard type.

CATANIA 1895

BUILT 1895 by D. & W. Henderson, Partick, Glasgow   DESIGNED G.L. Watson
LOA 203 feet   BEAM 26 feet 6 inches   GROSS TONS 588
ENGINE 4-cylinder triple-expansion 18, $28\frac{1}{2}$, 32, 32 × 27 stroke
RIG screw schooner

Catania, c.1899, lying at anchor off Cowes, is another excellent example of G.L. Watson's work. It is unusual to see her name painted on her bow as most steam yachts had their names painted on the stern. She also sports a fine figure-head and filigree on her bows.

Built for the Duke of Sutherland, the *Catania* stayed in his service until 1915, when she was hired by the Royal Navy for war service. Fitted with two 6-pounder guns, *Catania* served with the Royal Navy until February 1919, when she was sold to the Sun Shipping Co. Ltd, of Glasgow. By 1922, *Catania* had been sold to a Greek shipping company to be used as a cargo boat, having been renamed *Moschanthi Toqia*.

GRETA 1895

BUILT 1895 by J. Scott & Co., Greenock LOA 152 feet BEAM 22 feet
ENGINE triple expansion $12\frac{1}{4}$, 20, 32 × 26, by J. Scott & Co., Greenock
MATERIAL steel GROSS TONS 232

Greta was built for John Scott, of J. Scott & Co., Greenock, who owned her until 1897. After this she had a total of fourteen other owners until 1949, when she was scrapped. *Greta* was renamed four times during her career: *Zaranda* 1898, *Dorothy* 1910, *Zaranda* 1922 and *Devonia* in 1924.

Greta is a little unusual as she is rigged with square sails on her foremast. In the main this practice had largely died out in steam yachts by 1890. Also, *Greta*'s life boats are painted white; the normal practice was to have them varnished, although I feel the effect is quite pleasing.

The *Lady Sophia* being refitted possibly at J. Samuel White's yard. Bilge keels can clearly be seen, having been fitted to stop her rolling at sea. On the stocks beside her is the yacht *Lucille*.

Children on board their father's steam yacht *Marjorie*, 1895. *Marjorie* was later named *Galatea*, and served her owners faithfully for many years.

LADY SOPHIA 1895

BUILT 1895 by Ramage & Ferguson Ltd., Leith
DESIGNED St Clare Byrne   LOA 150 feet 7 inches
BEAM 21 ft   MATERIAL steel
ENGINE triple expansion 13, 21, 34 × 22,
by Ramage & Ferguson
GROSS TONS 203

The *Lady Sophia*, c.1900, on the stocks, possibly at J. Samuel White's yard at Cowes. At some time earlier in her life she had bilge keels fitted to her: one can only surmise that she had tended to roll at sea. It is interesting to note the lack of equipment in the yard, compared with that used today. Nevertheless, excellent work was carried out by the South Coast Shipbuilders. *Lady Sophia* had only one owner in England before she was bought by the French Government. Her owner, Mr W.H. Robertson, kept her from the time of completion in 1895 until his death in February 1918. By 1927 she had been purchased by Italian interests. Her other names were *Luciole*, *Nino Pittaluga* and *Silvia Onorato*.

Left The steam yacht *Gladys* having her bottom scraped and repainted at East Cowes, 1895.

SCUD 1896

BUILT 1896 by J. Samuel White of Cowes
LOA 84 feet 6 inches MATERIAL steel
ENGINE compound – 3 cylinder

This most interesting yacht was built for A.H. Wood, who owned her until 1898, when she was bought by W.C.S. Connall, in 1899. By 1902 she had once again been sold. Her new owner Thakur Sahib kept her until 1909, when she was wrecked. Built on the lines of the naval torpedo boat, she was designed for high speed. The photograph above was taken in 1905.

PUFFIN 1896

BUILT 1896 by Simpson Strickland & Co., Dartmouth
LOA 48 feet MATERIAL wood BEAM 9 feet

This official photograph was taken by Kirk of Cowes in 1896 and shows the steam launch *Puffin* at anchor in Cowes, with her owner H.T. Curtis. The vessel, built of teak, survived until 1906.

LATONA 1896

BUILT 1896 by Day Summers & Co., Northam, Southampton
LOA 144 feet 8 inches   BEAM 20 feet 2 inches   DRAFT 11 feet 7 inches
MATERIAL steel   ENGINE compound 18, 35 × 24, by Day Summers
GROSS TONS 184

Latona underweigh off the Isle of Wight, 1910. Carrying a crew of ten, she makes an imposing picture with her owners under her aft awning. *Latona* had had three owners before she was sold to Greece in 1919: Baron Knoop, 1896–1901; Sir I. W. Ellis, 1902–4; and I.G. Fay & Co., of Southampton, 1906–15.

HECLA 1896

BUILT 1896 by J. Samuel White of Cowes   LOA 80 feet   BREADTH 15 feet
DEPTH 9 feet 3 inches   TONS 85

Hecla was built for Mr R.A. Johnston, who owned her until 1898. She then had three other owners until 1915, when she was converted to a fishing vessel.

Hecla is seen here at Cowes in June 1896, with her owner R.A. Johnston on board.

MALLARD 1897

BUILT 1897 by J. Reid & Co., Glasgow   LOA 94 feet
GROSS TONS 89   MATERIAL steel

This fine little yacht had many owners, most of whom were based in Glasgow. Her name was changed several times: *Allah Karin* in 1908, *Mallard* in 1911 and *Maxtino* by 1938.

She is seen here after being reconditioned by J. Samuel White in the early 'thirties.

MAYFLOWER 1897

DESIGNER G.L. Watson   BUILT 1897 by J.G. Thompson
LOA 318 feet   BEAM 36 feet 5 inches
DRAUGHT 18 feet 5 inches   GROSS TONS 1779
MATERIAL steel   RIG twin screw schooner
ENGINE two TE $22\frac{1}{2} \times 38 \times 40 \times 40/27$ each
SPEED $16\frac{1}{2}$ knots

Mayflower and her two sister steam yachts *Varuna* and *Nanma* were designed to luxurious standards by England's foremost designer of steam yachts, G.L. Watson, and must be classed as the thoroughbreds of the steam yacht world in the late 1890s. Her first owner, Ogden Goelet, owned the *Mayflower* for less than a year due to his untimely death in 1898, after which the United States Navy purchased her for use in the United States Coast Guard Service. After being refitted and armed, she joined other naval ships in the blockading of Cuba and Havana during the Spanish-American War.

In 1902 *Mayflower* was recommissioned as the presidential yacht for the President of the United States, and in this capacity she served for many years. It was whilst on board the *Mayflower* that President Theodore Roosevelt signed the treaty ending the Russo-Japanese War.

During the First World War, *Mayflower* served on the east coast of the United States. After the war she continued to be used as the President's yacht until 1929, when it was found that the *Mayflower* was too expensive to run, consequently she was withdrawn from the commissioned list.

In 1931 she was badly damaged by fire whilst being laid up at the Philadelphia Navy Dock Yard, but by the mid-'thirties she had been sold by the Navy, refitted and sold once again to South American trade interests and renamed SS *Butte*. With the outbreak of the Second World War, shipping losses were heavy, due to enemy action at sea, and in 1942 the SS *Butte* was once again commissioned into the United States Coast Guard Service and renamed *Mayflower*. The *Mayflower* continued with the Coast Guard until 1946, when she was sold to Panamanian interests and renamed the *Mala*.

The illustration above shows the *Mayflower* underweigh off Cowes, c.1898.

APHRODITE 1898

DESIGNER Chas Hanscom
BUILT 1898 by Bath Iron Works, United States of America
LOA 302 feet 6 inches   BEAM 35 feet 6 inches
DRAUGHT 16 feet   GROSS TONS 1148
MATERIAL steel   ENGINE one TE 28 × 43 × 70/36

Originally built for Oliver Payne, who owned *Aphrodite* from 1898 to 1917, she was one of the largest steam yachts to have been built in the United States.

When the United States came into the First World War, Oliver Payne turned the yacht over to the US Navy in 1917. In company with *Corsair, Norma* and *Kanahwa*, she became part of the Breton Patrol in which she served until 1919. From 1919 to 1920 *Aphrodite* was owned by Harry Payne Whitney, who sold her to commercial interests in Greece. She is seen in Cowes Roads in 1903.

A society wedding of the yachting fraternity 1898, at Cowes, Isle of Wight. The happy couple even had their floral arrangements made in the form of a ship's bell.

Top right Cowes Week 1898. Underweigh amidst a cluster of yachts is a little steam yacht making for the Solent. Her owner is a member of the Royal Yacht Squadron. In the background, on the right-hand side, can be seen the royal yacht *Victoria & Albert II*.

Right A busy scene at Cowes Week, 1899. With the popularity of Cowes as a yachting centre, and that of Cowes Week, moorings were difficult to obtain. At the height of the season it was hard to see the water for the steam and sailing yachts.

51

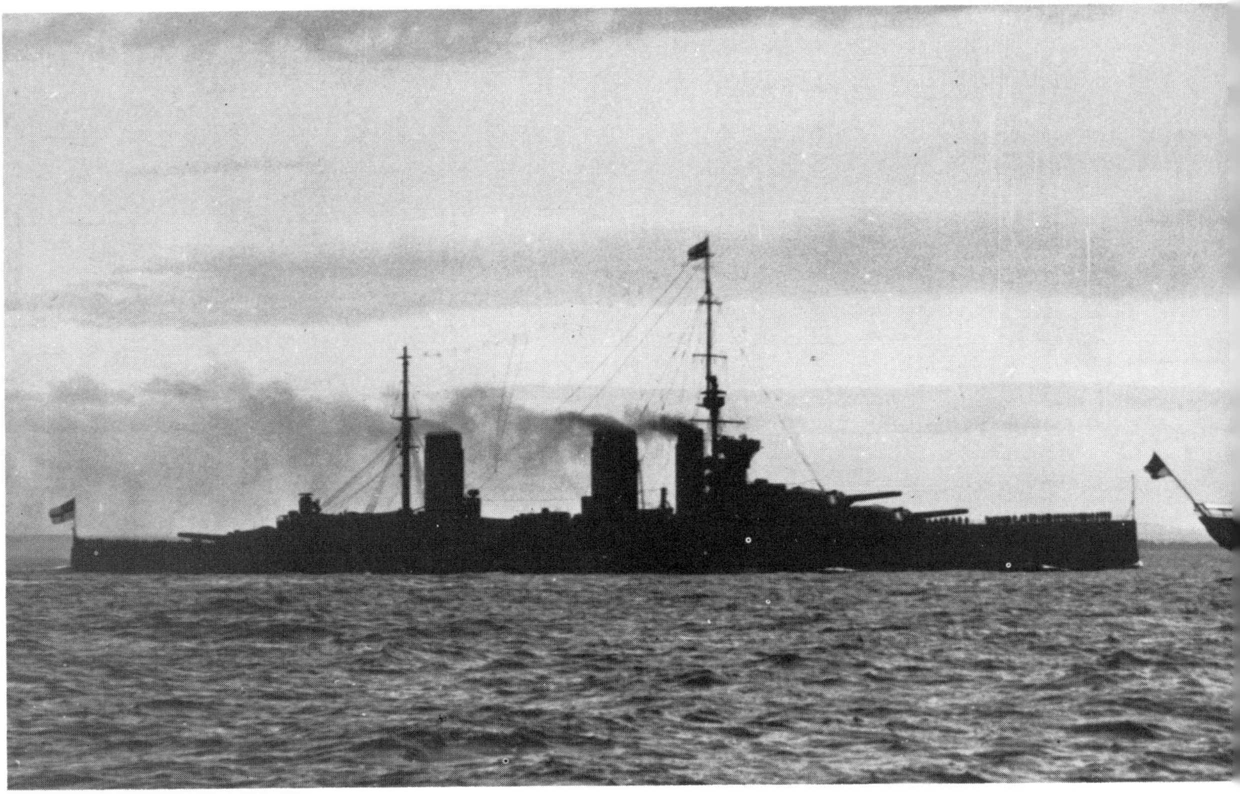

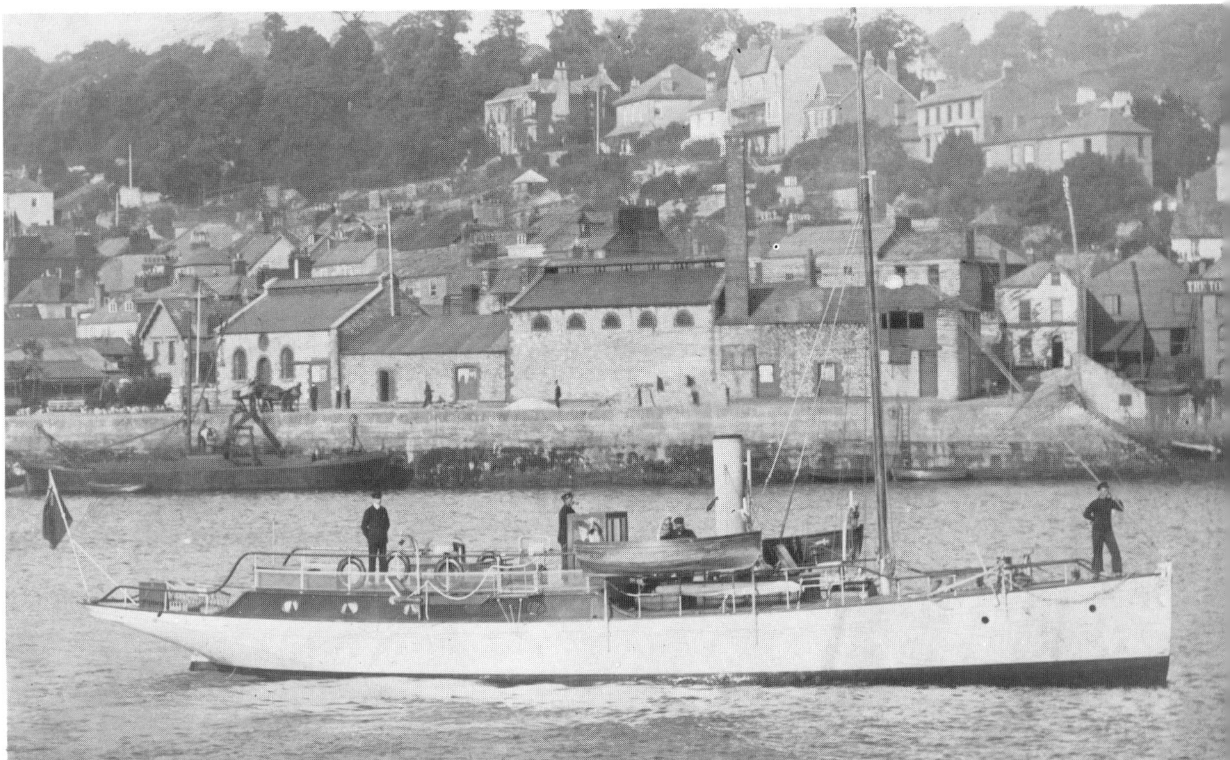

VICTORIA & ALBERT III 1899

DESIGNER Sir William White   BUILT 1899
MATERIAL steel   LOA 430 feet   BEAM 50 feet
DRAFT 17 feet   GROSS TONS 5500
RIG twin screw schooner
ENGINE two TE $26\frac{1}{2} \times 44\frac{1}{2} \times 53 \times 53/39$ each
KNOTS 20.7

CHRYSOPRASE 1899

BUILT 1899 by W. White & Sons
LOA 61 feet 7 inches   BEAM 11 feet 6 inches
ENGINES $8-16\frac{1}{2} \times 10$ GROSS TONS 34

Chrysoprase is interesting because of the fact that she has survived until the present day, although not in her original form, as she is now a houseboat on the Thames at Chertsey. Originally built for Mr H. Emmons of Southampton, who owned her from 1899 to 1900. By 1904 she had been bought by Harry Brickwood, Director of Brickwoods Brewery. She stayed in that family for many years, being converted into a houseboat in 1922.

The photograph was taken off Torbay in September 1903.

It was said that Queen Victoria disliked the *V & A III* to the extent that it was never used by Her Majesty. Perhaps this was due to an accident in dry dock which delayed the commissioning of the *V & A III* until 1901. Also, Queen Victoria preferred paddle to screw. In any event, *V & A III* became the official royal yacht of King Edward VII. She served five monarchs as the royal yacht in a total of fifty-five years and during that period made several trips abroad, including to the Baltic, Russia and the Mediterranean. In home waters she was used extensively for Fleet Reviews and Cowes Week. In 1954 the royal yacht *Britannia* was completed and commissioned, replacing the old *V & A III* which was scrapped in the following year.

The photograph above, taken on 26 July 1914, shows the royal yacht with her guard ship off the Nab Buoy.

A hive of activity at the fitting shop of J. Samuel White's Works at East Cowes. Many people were employed in the shipyards at Cowes, where the quality of workmanship was second to none.

The machine shop at J. Samuel White's Works at East Cowes.

Messrs S.E. Saunders' marine works on the banks of the river Medina, East Cowes, *c.*1900.

LADY FORREST

Although *Lady Forrest* was not used around the British coast, she is similar to the type that was. She was fitted with an ordinary screw propeller, unlike previous life boats, which were either fitted with water turbine or a screw propeller in a cavity in the vessel's hull under the cockpit. The length overall was 56 feet, with an extreme breadth of 15 feet 9 inches. She attained a speed over the measured mile of 10.1 knots with all weight on board.

Lady Forrest is seen here on trials before being shipped to Fremantle to take up her duties.

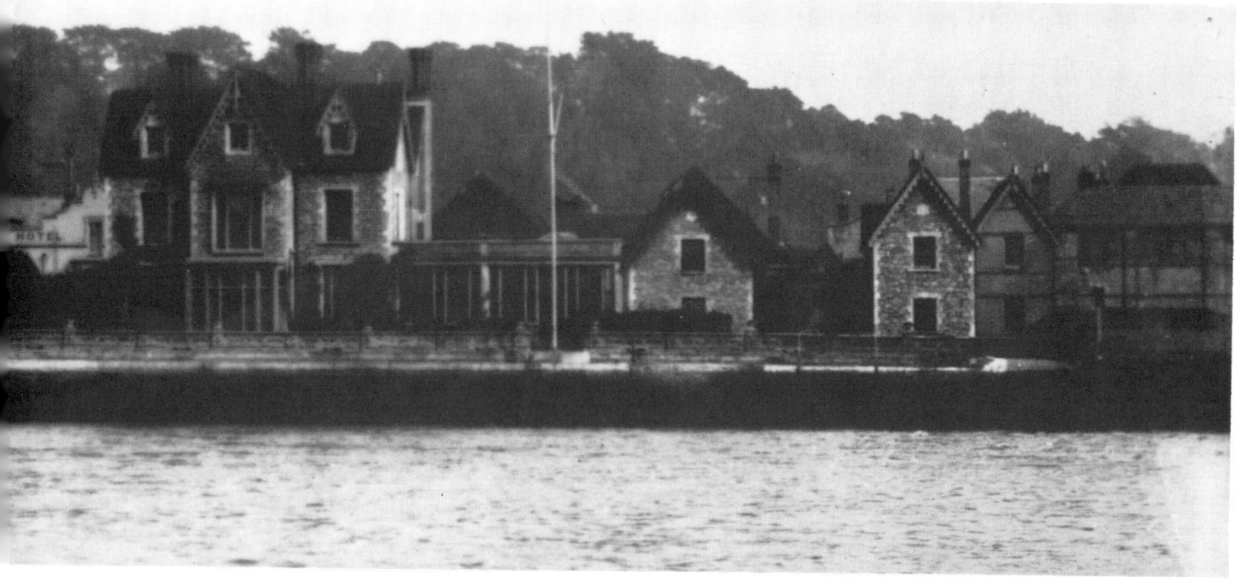
HOTEL

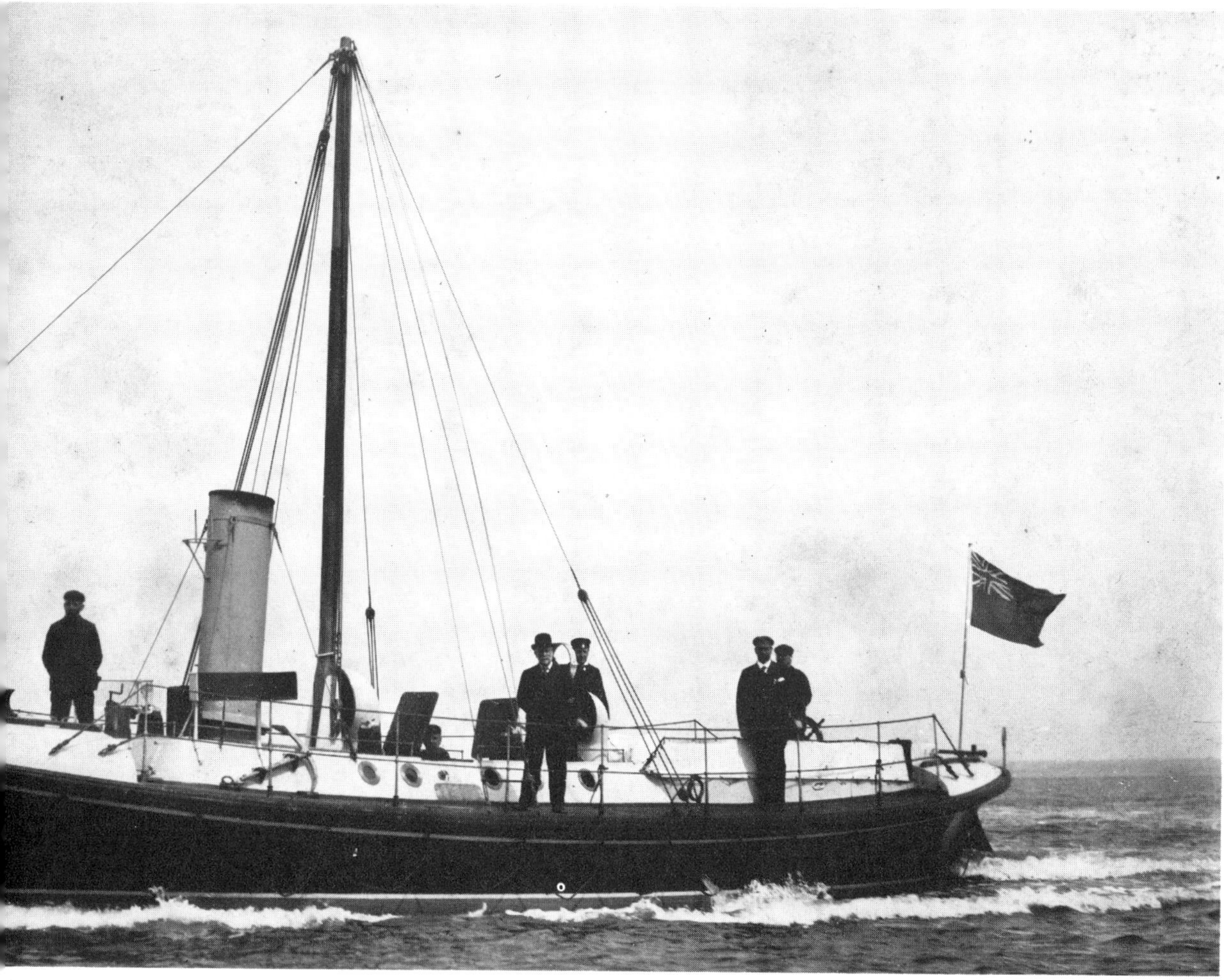

ZAIDA 1900

BUILT by J.S. White of Cowes LOA 149 feet RIG twin screw schooner
ENGINE TE GROSS TONS 255

The *Zaida* was typical of the high standard of craftsmanship produced by J.S. White of Cowes. Showing neat and compact lines, she must have looked beautiful at sea, with her gleaming white hull and gold painted filigree at her bows.

Zaida was first owned by Mr A. Shuttleworth of Cowes, from 1900 to 1901. She was then sold to the Earl of Rosebery KG, KT, who was Commodore of the Royal Fowey Yacht Club, and remained in his service until 1915. On 26 May 1916, *Zaida* was hired by the Royal Navy as an auxiliary patrol vessel.

Having been armed with two 6-pounder guns, *Zaida* served in the eastern Mediterranean until 17 August 1916, when she was attacked by gunfire from the German submarine *U.35* near Alexandria. During the ensuing attack, the *Zaida* was hit and later sank.

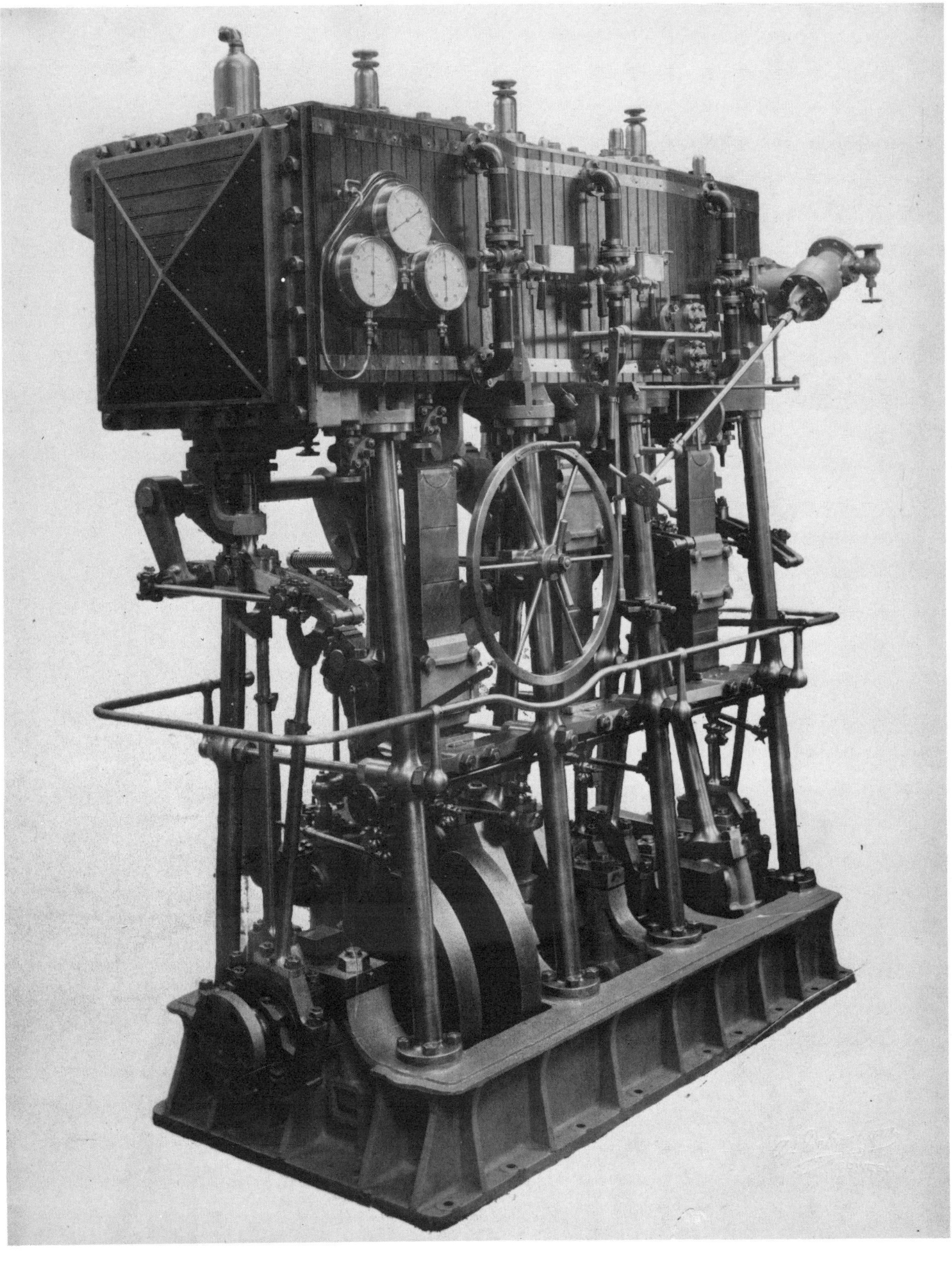

Left The lounge of the steam yacht *Zaida*, 1900. The wood fittings were of teak, with wall-to-wall carpeting and well-upholstered chairs creating an atmosphere of opulence and comfort.

Left The *Zaida*'s writing room, 1900. Note the fine wooden pillars and the mantle clock. The clock was presented to the *Zaida*'s first owner, Mr A. Shuttleworth, by the Royal Yacht Squadron.

Opposite page Triple-expansion engine 11.17 × 26 × 20 designed and built by J.S. White of Cowes, before being installed in the steam yacht *Zaida*, 1900.

Triple-expansion engines became standard in most yachts by the 1890s. With the development of high-pressure boilers, it became possible to instal the compound engine, using three cylinders of high, medium and low pressure. The working action gave less vibration, more speed and greater economy of steam.

The steam yacht *Zaida*, taken in 1900 by A. Debenham, was possibly one of the most beautiful steam yachts to emerge from J. Samuel White's yard at the turn of the century.

Below

BUILT *c.*1900 by the Liquid Fuel Engineering Co., Cowes, Isle of Wight
LOA 70 feet   RIG twin screw steam yacht
ENGINE 100–1 h.p.   SPEED 12 knots   MATERIAL wood

Unfortunately it is difficult to identify this fine little craft as no name could be found in the records. Looking very much like a large launch, it is almost certain that the photograph below shows this yacht on sea trials in the Solent. Her owner at that time was a Captain Fripain.

EDA

BUILT c.1900 by the Liquid Fuel Engineering Co., Cowes, Isle of Wight

The *Eda* was built at Cowes by the Liquid Fuel Engineering Co., who traded under the name of LIFU. Apart from having yacht-building works at Cowes, they also had engineering works at Bitterne Park, Southampton and at Poole Quay, Dorset. LIFU were one of the first firms to introduce petroleum as a form of fuel in place of coal or wood. In their catalogue of 1910, they give an account of where petroleum could be obtained:

> The recent discoveries of fresh petroleum fields on the European as well as on the American Continent, the prompt measures taken to lay pipe lines to the sea coast where necessary, and the construction of tank steamers on the newest and most improved principles, have combined to place mineral oil in almost every port in the world, where it can be bought at prices ranging between 2d. and 6d. per gallon.

This particular launch was not typical of the launches advertised in the LIFU catalogue, as the *Eda*'s length was 36 feet, the standard lengths being 16, 18, 20, 22, 23 feet 6 inches, 25, 27, 30, 35 and 45 feet. In some cases, the cabins were detachable, no doubt the better to enjoy the hot summer days at sea, but with the *Eda* this was not the case, as her cabin was fixed.

The above illustration shows the *Eda* flying the pennant of the Royal Victoria Yacht Club, established in Ryde, Isle of Wight, in 1844. In the background centre can be seen the royal yacht *Elfen*. *Elfen*, a paddle yacht, was built in 1849 and had an overall length of 112 feet 3 inches, her beam being 25 feet over her paddles. Commonly known as 'the milk boat', she crossed between Portsmouth and Southampton from Osborne House, on the Isle of Wight, daily, with royal despatches and papers.

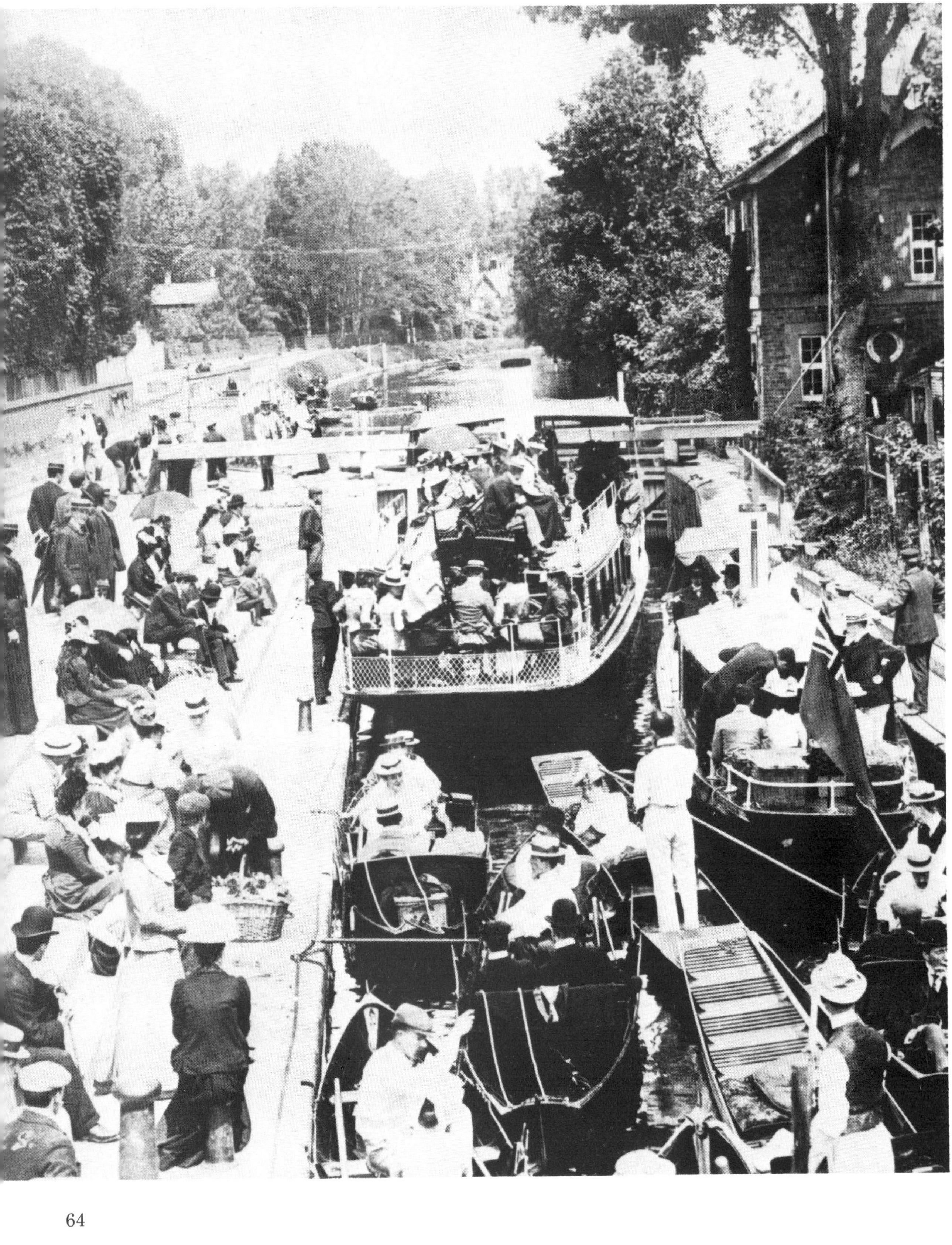

Steam launches were in great demand on the Thames to convey holiday makers on trips along the river. The illustration was taken at Boulter's Lock in 1900.

Left A view of Boulter's Lock in 1900. All seems ready for the afternoon trip up the river on the steam launch.

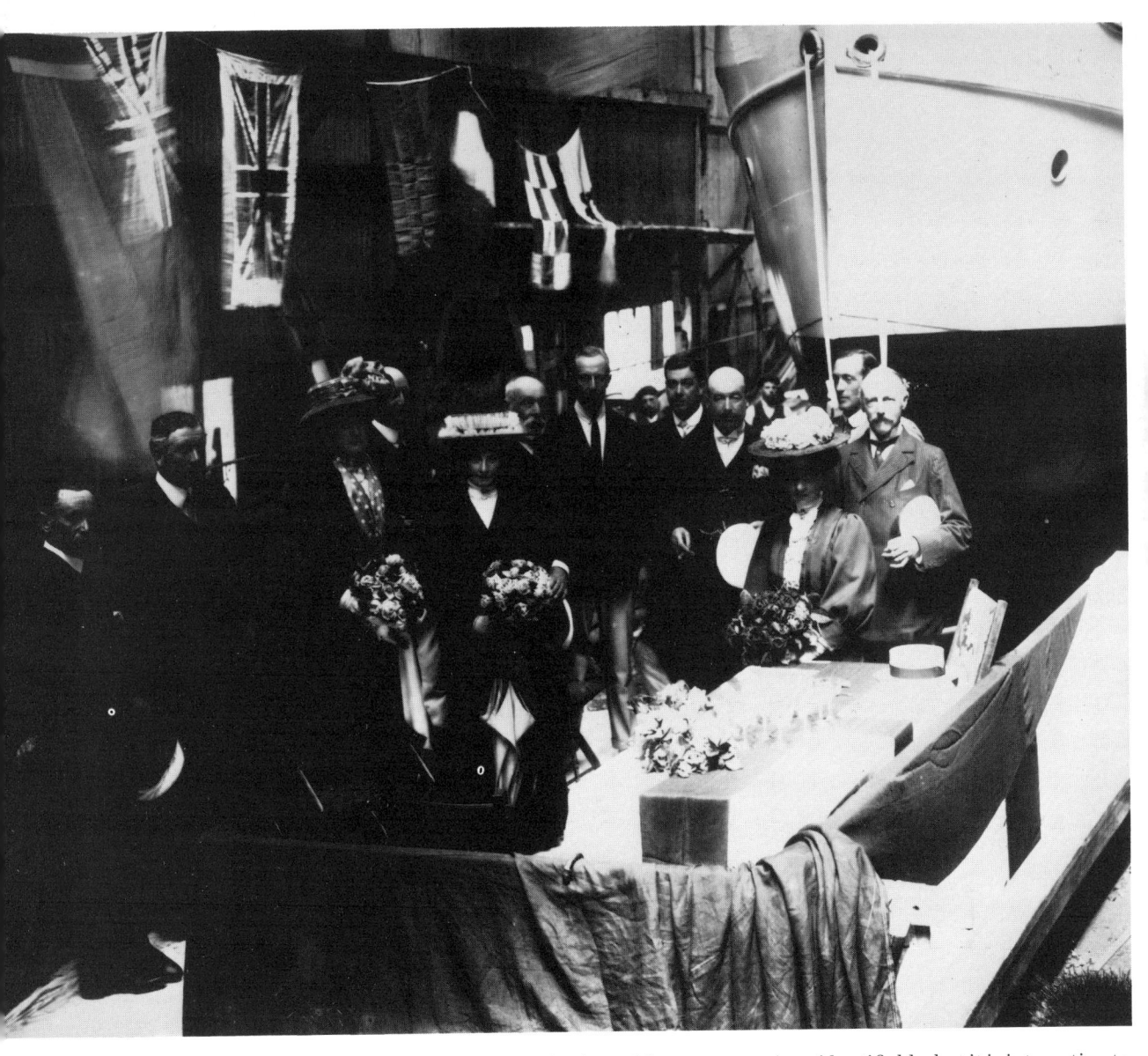

The yacht in this launching ceremony is unidentifiable, but it is interesting to note the dress at a society launching. The gentleman on the extreme right is Sir William Portal, a then well-known yachtsman.

Sir William owned the yacht *Star of India* (545 gross tons), which was originally named *Valfreyia*. One of the many owners of the *Star of India* was the Maharaiah of Nawanagar, who was better known as Ranjit Sinjhi, the cricketer.

The yacht basin at Trouville, Deauville, France, c.1901. The French ports were very popular with the yachting fraternity. In the illustration a fine collection of thoroughbred steam yachts are anchored in their berths.

LENSAHN 1901

BUILT 1901 by A.G. Howaldtswerke of Kiel, Germany LOA 164 feet
GROSS TONS 513 ENGINE two TE $12\frac{1}{2} \times 19\frac{1}{2}$, $31\frac{1}{2} \times 18\frac{1}{2}$ stroke
RIG twin screw schooner

Built for the Grand Duke Van Oldenberg, the imperial yacht *Lensahn* is not unlike the imperial yacht *Hohenzollern*, built in 1893; although *Lensahn*'s hull is different, the design of her superstructure is rather similar.

Lensahn is seen here anchored off Cowes. The design of her funnels is bell-shaped, although with British royal yachts, this style was more distinctive.

A charming society photograph taken by Kirk of Cowes in 1901, on board the owner's steam yacht. Note the fine embroidery on the coat of the lady sitting on the right. It was all-important to be dressed in style at Cowes Week.

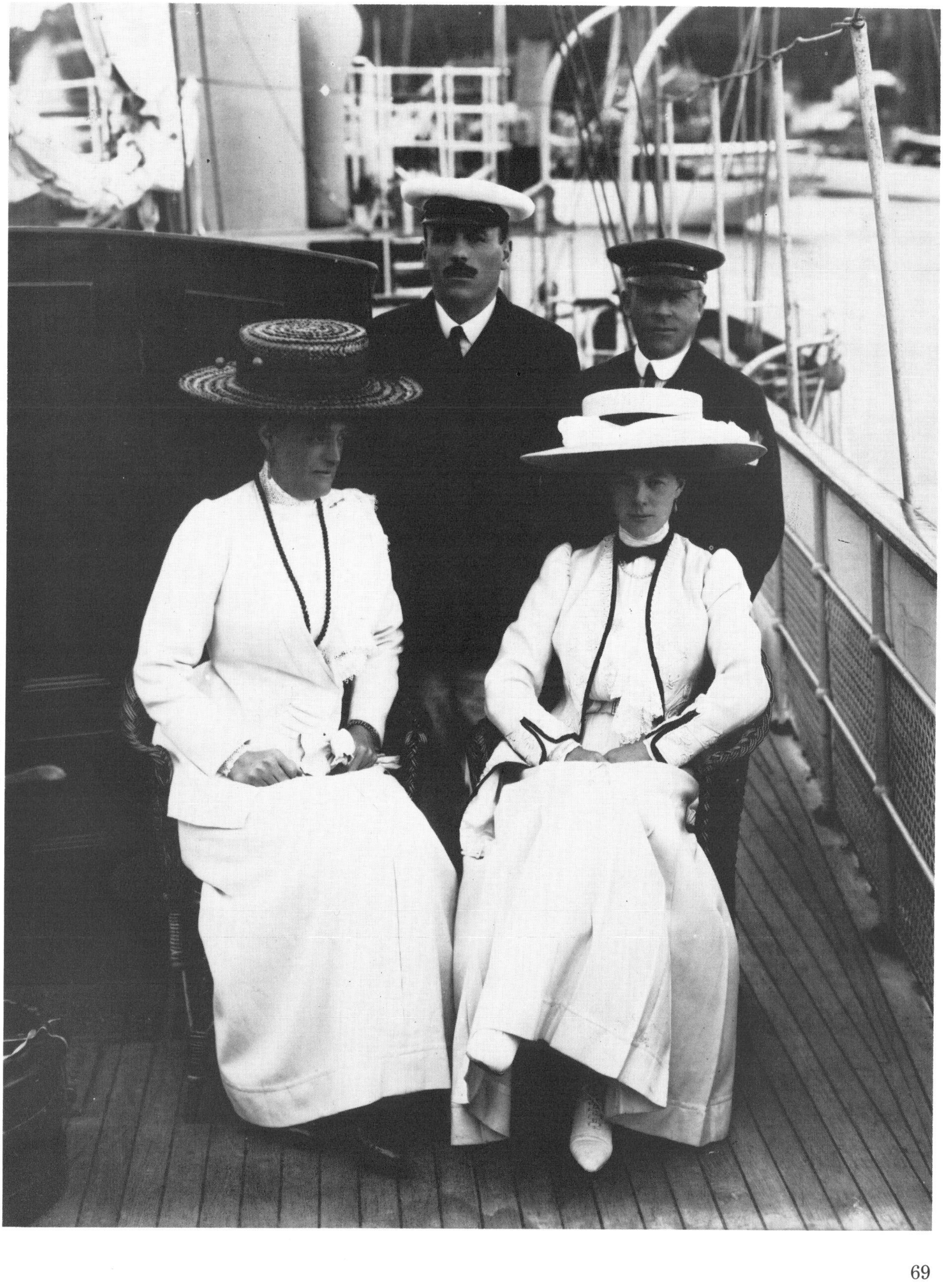

Striped blazers were the height of fashion in yachting circles in the early 1900s.

The royal yachts *Victoria & Albert II* and *Victoria & Albert III* at Cowes during Cowes Week c.1901. The royal yachts' guardship can be seen to the left, in the background.

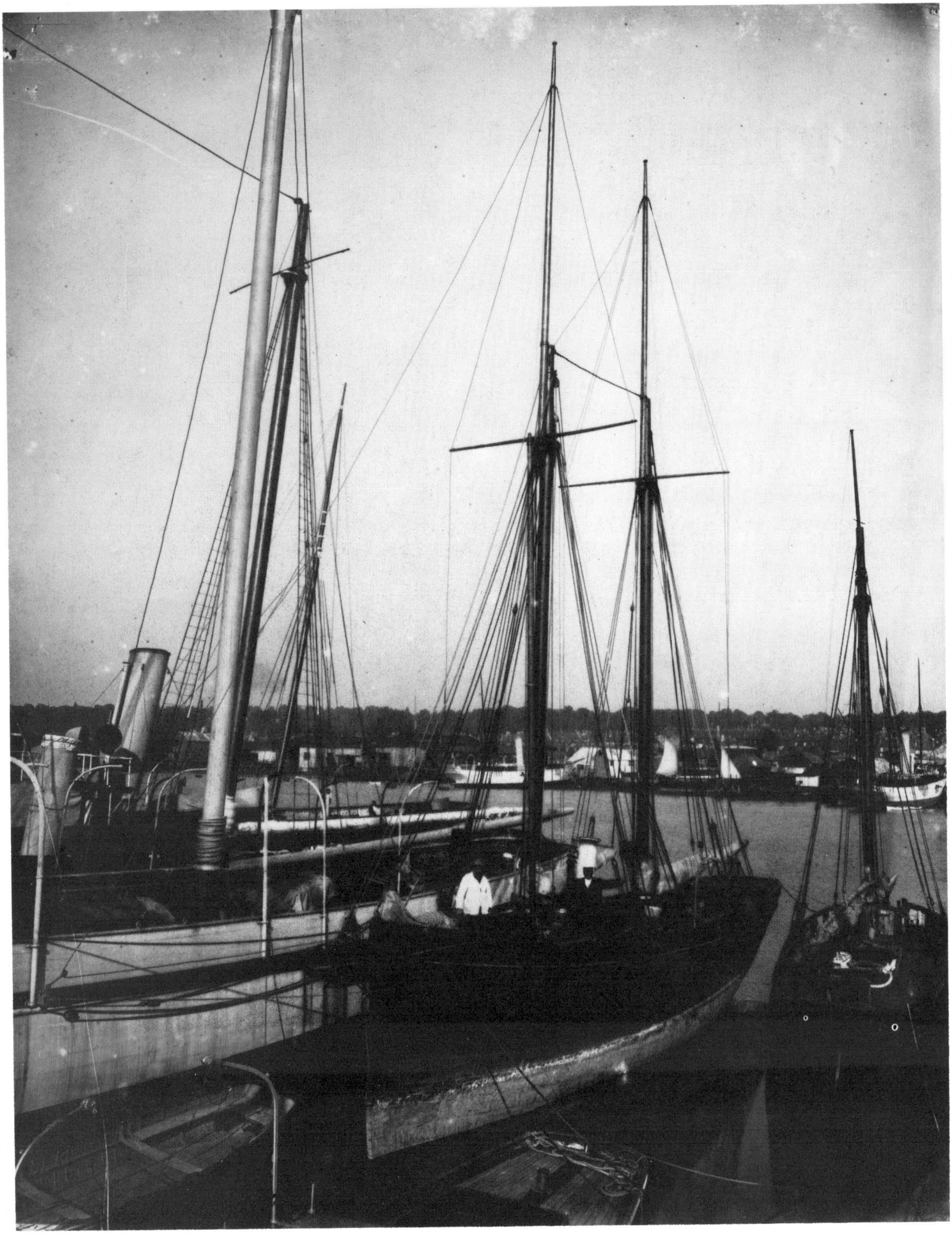

Steam yachts anchored at Weymouth during a regatta at the turn of the century. During the summer, owners would take their yachts from one regatta to another, throughout the season.

Left Steam and sailing yachts moored side by side on the banks of the river Medina. In the background may be seen the many shipyards which grew up along the Medina at East Cowes during the 1900s. The steam yacht in the left-hand corner has a canvas cover over her smoke stack to stop the rain seeping down into the boiler while the yacht was not being used.

Top left A photograph on board the steam yacht *Cuhona* of the yacht's officers. Wages ranged from First Mate's pay of £3 to £4 per week down to Boy Seaman, who earned 18 shillings per week.

Centre left The ship's crew on board the steam yacht *Cuhona*, taken at Cannes in the spring of 1902.

Left The steam yacht *Cuhona* getting underweigh at Spezzia, 1902.

Above The master of the steam yacht *Cuhona*, Captain Derham, pictured with the owner's family, in spring 1902. Captain Derham would have received in the region of £90–£100 per year in salary, plus free food and uniform.

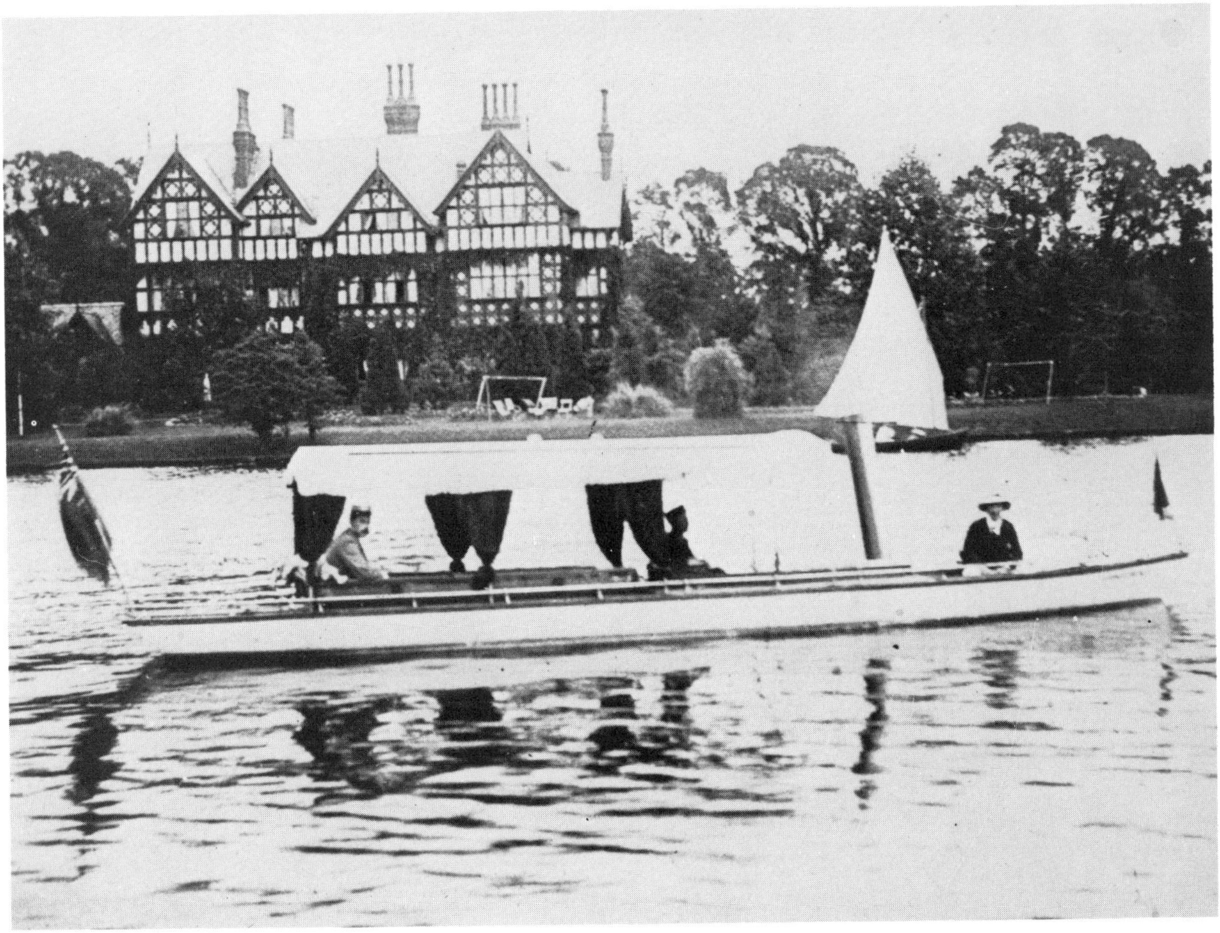

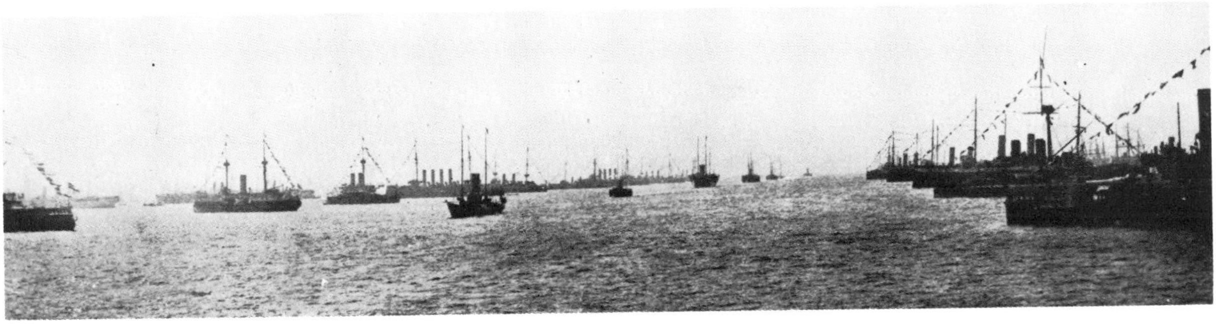

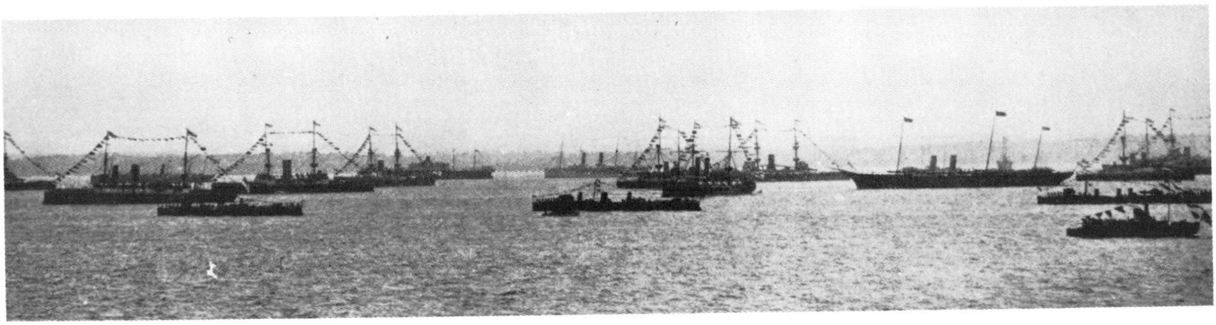

Opposite left The steam launch *Frivolities* on the Thames in 1902.

Left The Coronation Fleet Review of Edward VII. The King on board the royal yacht *Victoria & Albert III* is accompanied by the steam yachts *Irene*, *Alberta*, *Osborne*, *Enchantress* and *Fire Queen*.

CARMELA 1903

BUILT 1903 by Camper Nicholson, Gosport
DESIGNED C.E. Nicholson LOA 100 feet
BEAM 18½ feet DRAUGHT 11 feet
ENGINE compound 16, 30 × 18 inches by Day Summers & Co., Northam
GROSS TONS 123

Carmela is typical of the smaller steam yachts constructed in the early 1900s. In style, *Carmela* is not unlike the *Chrysoprase*. She was first owned by Mr C. Thellusson of Portsmouth, 1903–10, after which she was sold to a succession of owners until 1945 when her records cease.

Although *Carmela* was not requisitioned by the Royal Navy in the 1914–18 war, she did serve as an accommodation ship in the Second World War. Having been hired on 12 August 1941 by the Royal Navy, she was later bought by them in 1943 and eventually sold in 1945.

ADMIRALTY STEAM LAUNCH No. 63

Built by J. Samuel White for Osborne Naval College in 1903. Launches of this type were used to give cadets first-hand knowledge of using small boats at sea.

ENCHANTRESS 1903

BUILT 1903 by Harland & Wolff LOA 320 feet BEAM 40 feet
DRAUGHT 16 feet GROSS TONS 3370 MATERIAL steel
SPEED 18 knots H.P. 6400

Enchantress was built for the Royal Navy, to be used by the Lords of the Admiralty. Having been commissioned in 1904, she was used by the Admiralty until the outbreak of the First World War, when strangely she was laid up; one would have thought that she could have been used as a hospital ship. Nevertheless, *Enchantress* was not recommissioned until 1919. She carried on with her duties until 6 December 1934, when she was once again laid up. On 24 June 1935 she was sold for scrap.

Left A private steam launch on the Thames in the summer of 1903.

The dining room, seating eight people, decorated in natural wood.

Right The bridge of the *Ethleen*. Note the speaking tube to the engine room, also the powerful searchlight. The delightful wicker chairs were for family and friends, not the crew.

The steam yacht *Ethleen* was built in 1903 by the Palmer Shipbuilding and Iron Company of Jarrow.

The photographs were taken in 1928, when she was owned by the famous aircraft pilot, Claude Grahame-White.

The sitting room, with a coal fire and fitted carpets. In every case the steam yacht's interior was to be a home from home. Comfort was all important at sea.

Left The music room, wickerwork chairs, a piano and one's evening entertainment was ensured.

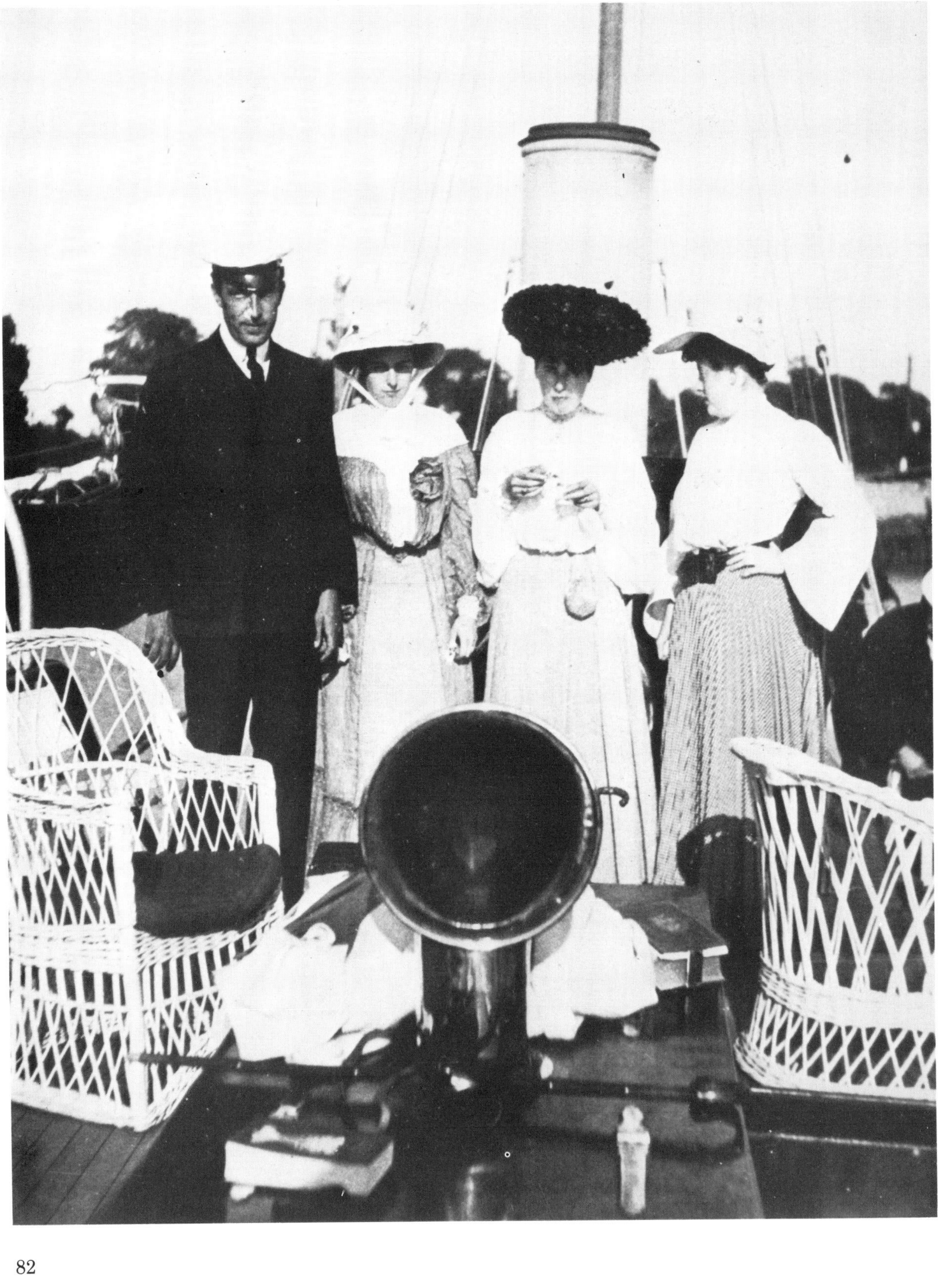

Below The captain and crew on board the steam yacht *Chrysoprase* in 1904.

Left Harry Brickwood, owner of the steam yacht *Chrysoprase*, entertaining friends on his yacht in the summer of 1904.

GELASMA 1904

BUILT 1904 by Gibbs & Co., Devon   LOA 88 feet   BEAM 20 feet
ENGINE compound $5\frac{3}{4}$, $11\frac{1}{2} \times 8$

Gelasma is unusual in that her appearance is like a trawler, but she was built as a yacht rather than being converted from a fishing vessel. It is possible that her first owner, Sir R.H. Williams Buckeley, had a liking for Brixham trawlers, as her styling is reminiscent of that type of craft. *Gelasma* had five owners: Sir R.H. Williams Buckeley, 1904–9; Dr A. Luling of Le Havre, 1910–11; Harold Marshall, 1912–14; Mrs J. Borthwick, 1915–18; and J.T. Leete, 1919. In 1920 *Gelasma* was sunk in a shipping accident.

The steam tug *Irishman* in Cowes harbour 1905. Steam tugs played an important part in the everyday life of a busy harbour, towing yachts to their moorings, helping in launching large steam yachts, and in many other duties. *Irishman* had the sad duty of towing King George V's great yacht *Britannia* out to her last mooring in Cowes Roads, to wait her final end, that of being scuttled. In King George's will, he stated that he wished *Britannia* to be scuttled rather than rot at her moorings. On the evening of 10 July 1936, two destroyers, HMS *Amazon* and HMS *Winchelsea*, escorted *Britannia* to a secret destination off the Isle of Wight. After laying explosives aboard her, the two destroyers manoeuvered to a safe position and at midnight detonated the charges. *Britannia* slipped slowly below the dark water, a fitting end for a supreme yacht. During her life she had won a total of 208 prizes in racing, crewed by a professional crew of thirty-five. She carried 17,000 feet of canvas when under sail.

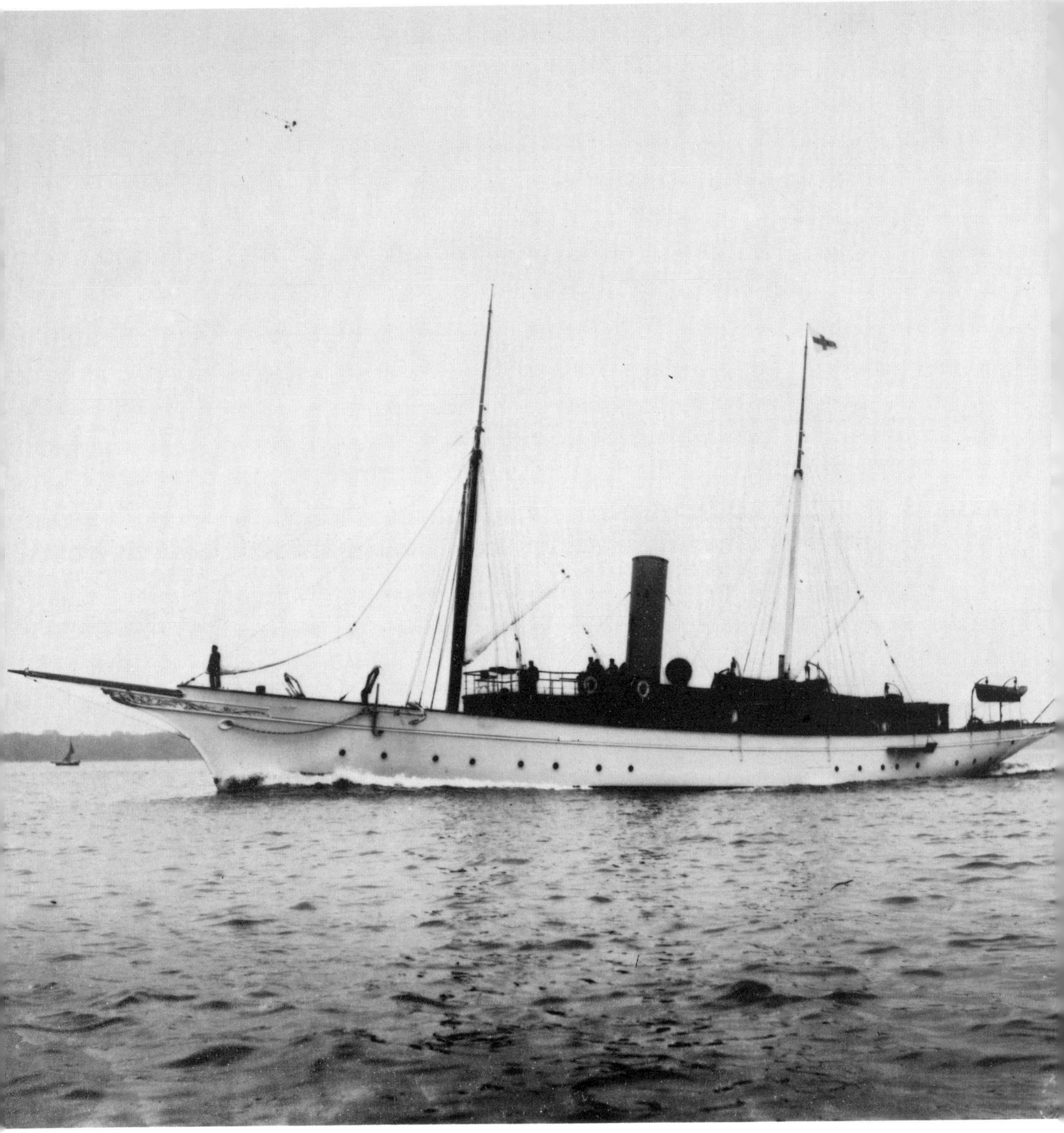

AGATHA 1905

BUILT 1905 by Day Summers, Northam, Southampton
LOA 164 feet 5 inches BEAM 24 feet 6 inches
DRAUGHT 12 feet 5 inches MATERIAL steel
ENGINE triple expansion 6cy. 11, $17\frac{1}{2} \times 28$–21

Agatha was turned out with every luxury befitting a fair-sized steam yacht of its time – including electric light: a luxury at sea, but even more so on land.

She was owned by Sir E.W. Greene until 1920, when she passed to the Earl of Inchcape who renamed her *Rover*. In 1930, *Rover* was purchased by H.W. Adams, who kept her until 1933 when she was bought by foreign interests. In 1914, *Agatha* was hired by the Royal Navy. Armed with one 12-pounder and one 6-pounder she was used as auxiliary patrol vessel until 1919, when she was returned to her former owner. Her other names were *Else* and *Nere-Naetza*.

Above The yacht *Gladys* getting up steam at Cowes c.1906.

The auxiliary steam yacht *Sunrise* after being converted to schooner rig by J.S. White in 1906.

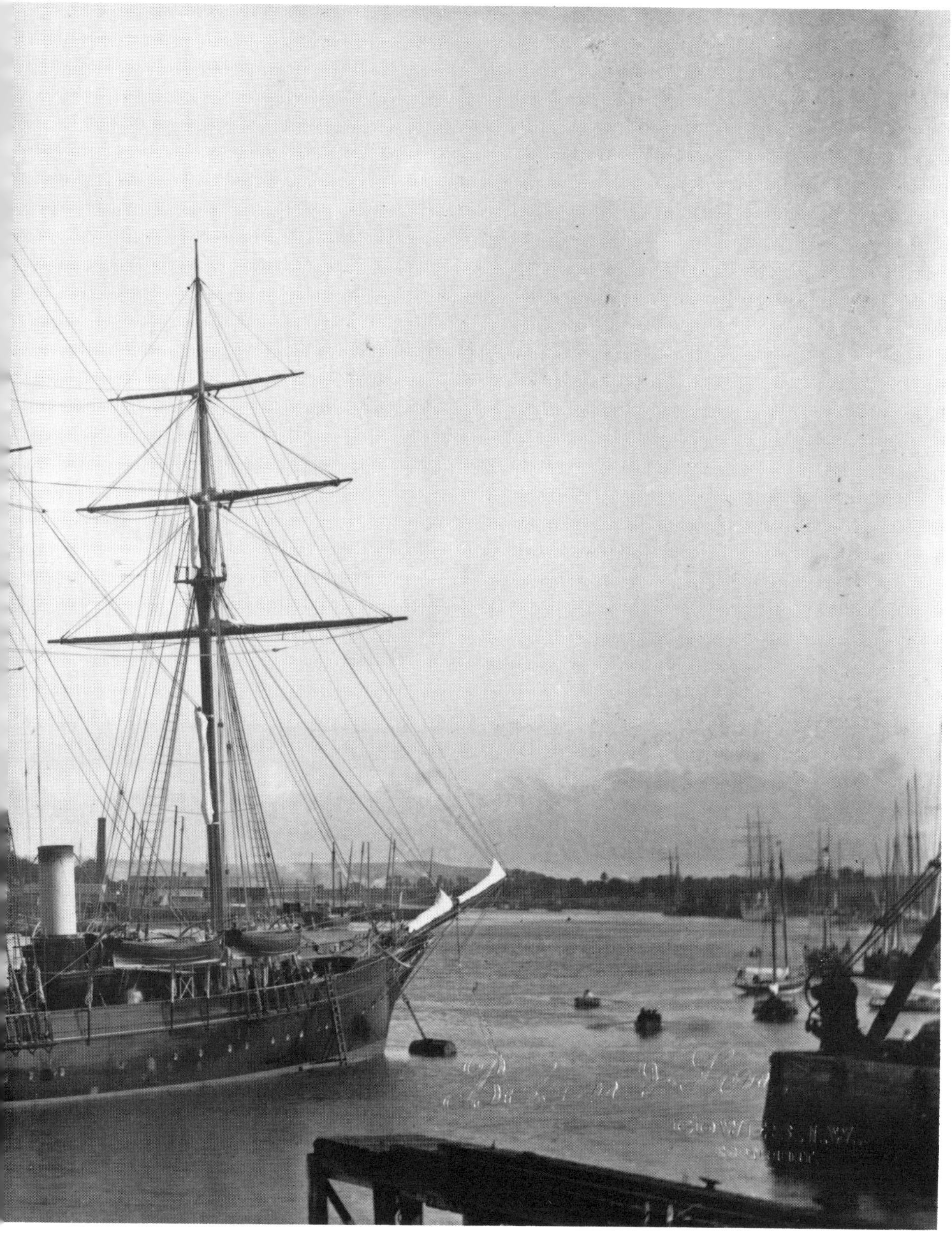

JADE 1906

BUILT 1906 by W. White of Cowes   LOA 54 feet
ENGINE compound 8, $16\frac{1}{4} \times 8$ inches stroke
MATERIAL composite iron or steel frames and wood planking

Jade is one of a few yachts to have survived until the present day. Originally owned by H. Emmons 1906–9, she passed through many hands until her final owner C.G. Wells sold her in 1970. Here her records end, but it is almost certain that after 1970 she was converted to a houseboat, as her engines were removed by 1949, no doubt to give extra accommodation space.

Jade, c.1910, is seen underweigh off Cowes and is flying the flag of the New York Yachting Club.

MEDUSA 1906

BUILT 1906 by Day Summers & Co., Northam, Southampton
LOA 199 feet 1 inch   BEAM 26 feet 1 inch   DRAUGHT 15 feet 3 inches
MATERIAL steel   ENGINE triple expansion 16, 26 × 42–27
RIG screw schooner

The steam yacht *Medusa* starting down the ways, 10 May 1906.

Medusa led a quiet life in British waters, having survived her service as an auxiliary patrol vessel in the First World War, but in 1939 she was once again hired by the Royal Navy for active service and renamed *Mollusc*. She carried out her duties until 17 March 1941, when she was attacked by enemy aircraft. During the engagement she was sunk off Blyth.

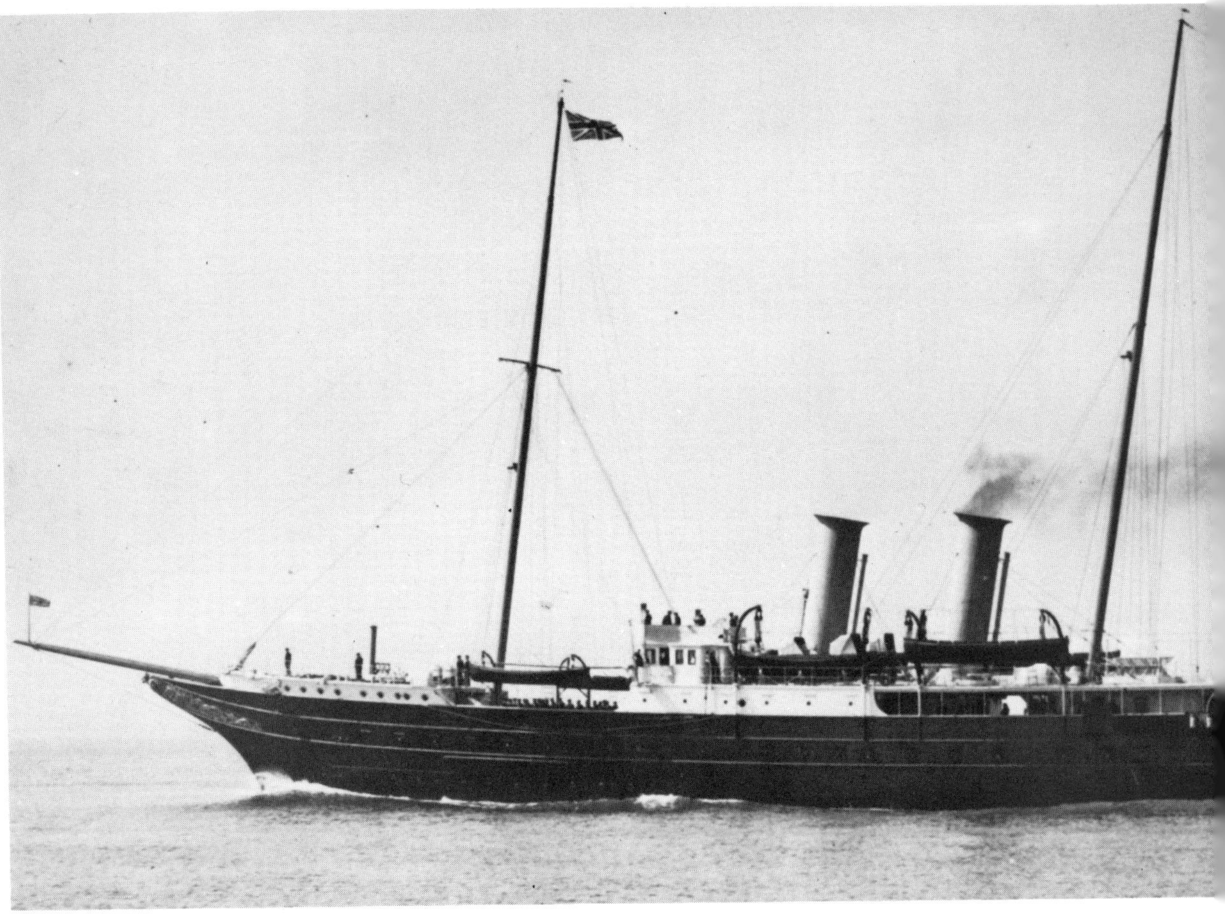

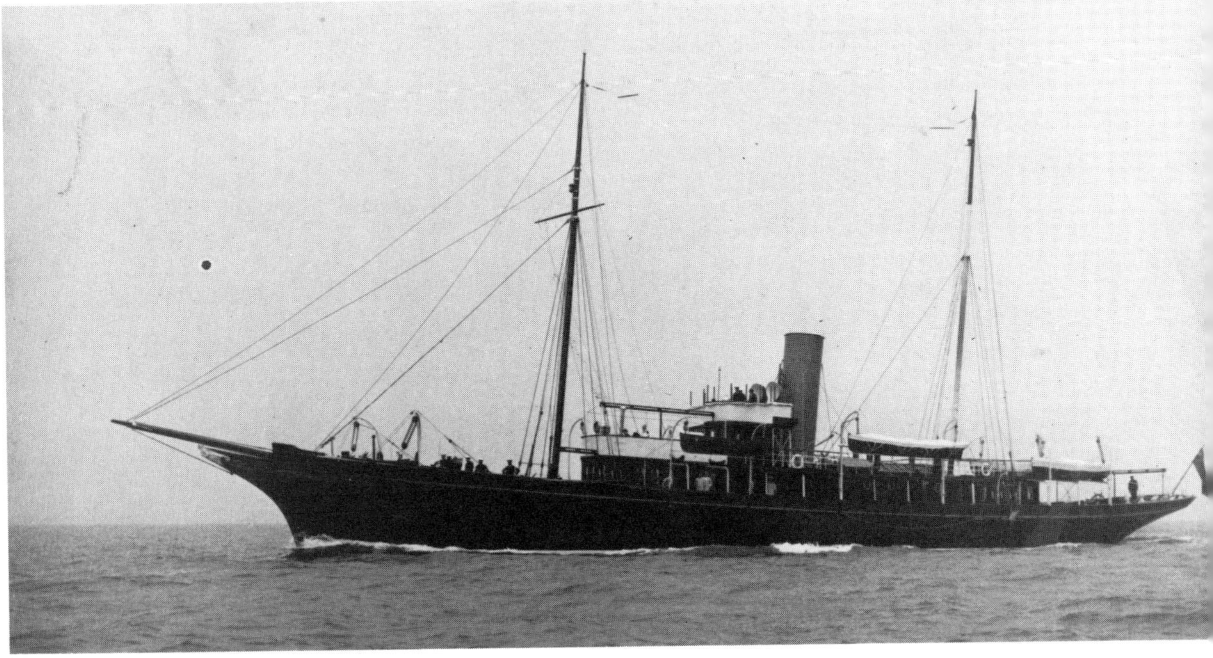

ALEXANDRA 1907

BUILT 1907 by A. & J. Inglis
LOA 325 feet   BEAM 40 feet   DRAUGHT 13 feet
GROSS TONS 1728   MATERIAL steel
ENGINE 3 Parsons turbines

The royal yacht *Alexandra* was built to replace the royal tender *Osborne*. She was used by King Edward VII, and in later years by King George V. Although similar to the *Victoria & Albert III*, she was considerably smaller in construction. Her main duties consisted of cross-channel trips. While in the service of George V, *Alexandra* was little used, as all the King's time was taken up with sailing his yacht *Britannia*. In 1925 she was sold to Norwegian commercial interests.

During the invasion of Norway in 1940 *Alexandra* was destroyed.

SYLVANA 1907

BUILT 1907   LOA 175 feet 9 inches
DRAUGHT 14 feet 3 inches   GROSS TONS 419
MATERIAL steel   BREADTH 24 feet 6 inches

Originally named the *Maid of Honour*, her first owner was W.K. Millar, 1907–12. She then had a long succession of owners up to 1952, when records cease. Whilst under the ownership of Miss K.A. Mackinnon, her name in 1923 was changed to *Sylvana*. During the First World War she served with the Royal Navy as an auxiliary patrol vessel and was armed with 12-pounder guns.

The picture was taken off the Isle of Wight in 1926.

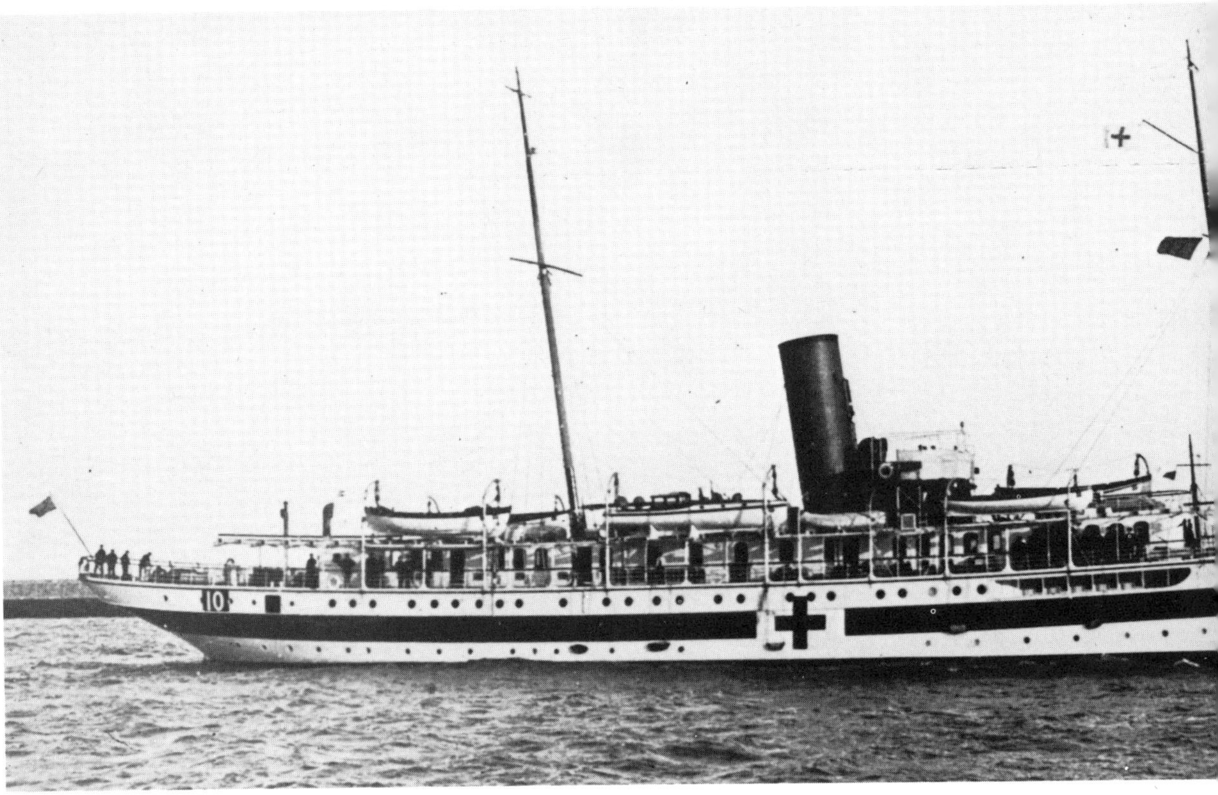
10

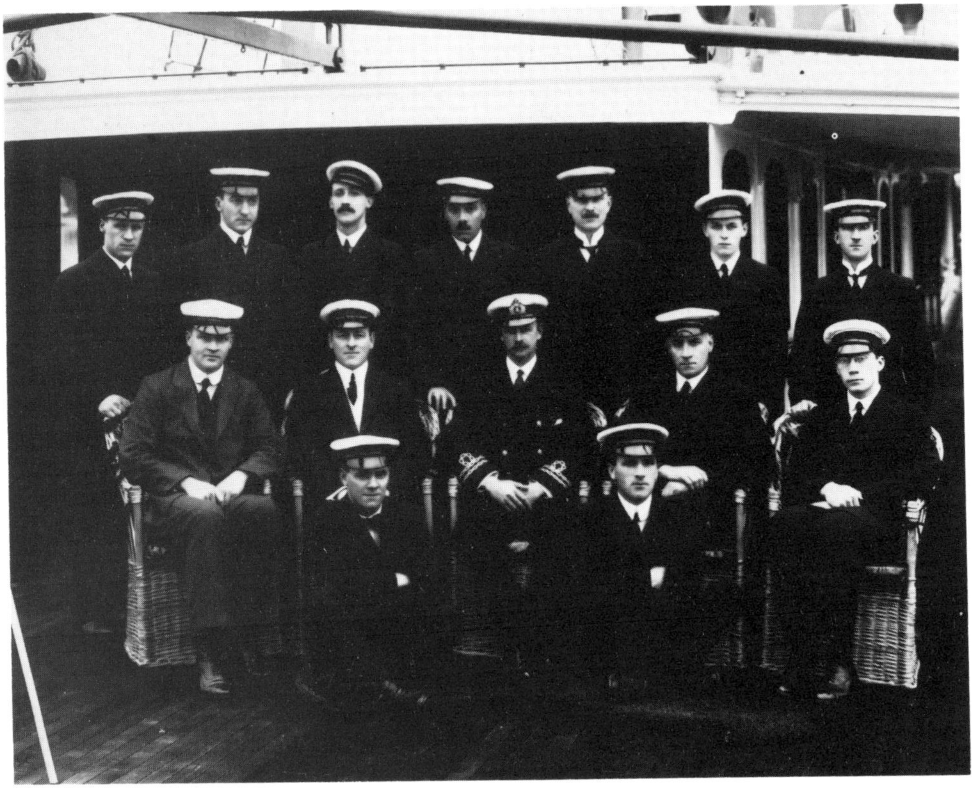

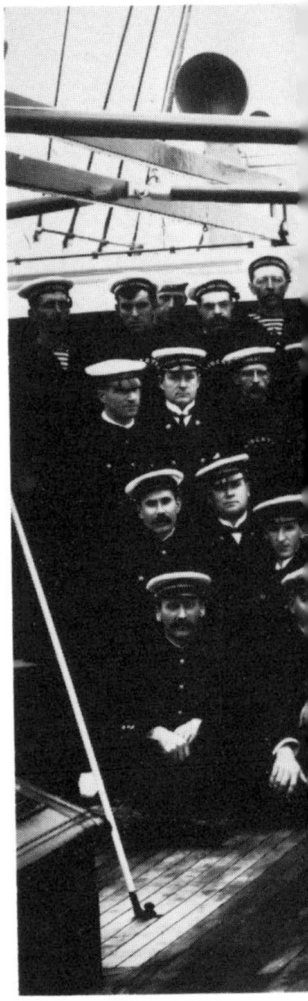

LIBERTY 1908

DESIGNERS G.L. Watson & Co.
BUILT 1908 by Ramage & Ferguson   LOA 304 feet
BEAM 36 feet 5 inches   DRAUGHT 16 feet
MATERIAL steel   ENGINE two TE 16 × 26 × 42/24 each

The steam yacht *Liberty* was originally built for Joseph Pulitzer, who owned her from 1908 to 1912. Pulitzer was nearly blind and ultra-sensitive to noise, and the designer, G.L. Watson, had to take this into account when he designed *Liberty*. Watson had the yacht heavily insulated against sound, whilst all sharp edges were reduced to curves, and instead of steps, gentle slopes were fitted. *Liberty*'s design was to become invaluable in later years, when she was hired as a hospital ship in the First World War. In 1912, James Ross became *Liberty*'s owner, and he re-named her *Glencairn*, but by 1914 she had been sold to Lord Tredegar, who reversed her name back to *Liberty*. Lord Tredegar kept *Liberty* for less than a year, before she was hired by the Royal Navy as an auxiliary patrol vessel.

On 1 September 1915, *Liberty* was re-assigned as a hospital ship, and served in this capacity until January 1919, when she was sold to Sir Robert and Lady Houston.

Lady Houston will always be associated with the *Liberty* as she used the yacht to display anti-Government signs hung between the masts, while at various regattas.

In 1937 *Liberty* was broken up.

In this photograph *Liberty* can be seen in her war-time colours as a hospital ship, c.1917.

Far left The medical staff of the hospital ship *Liberty*, 1915.

Left The war-time crew of the steam yacht *Liberty*, 1915. As *Liberty IV* in naval records, *Liberty* served as a hospital from 1915 to 1919.

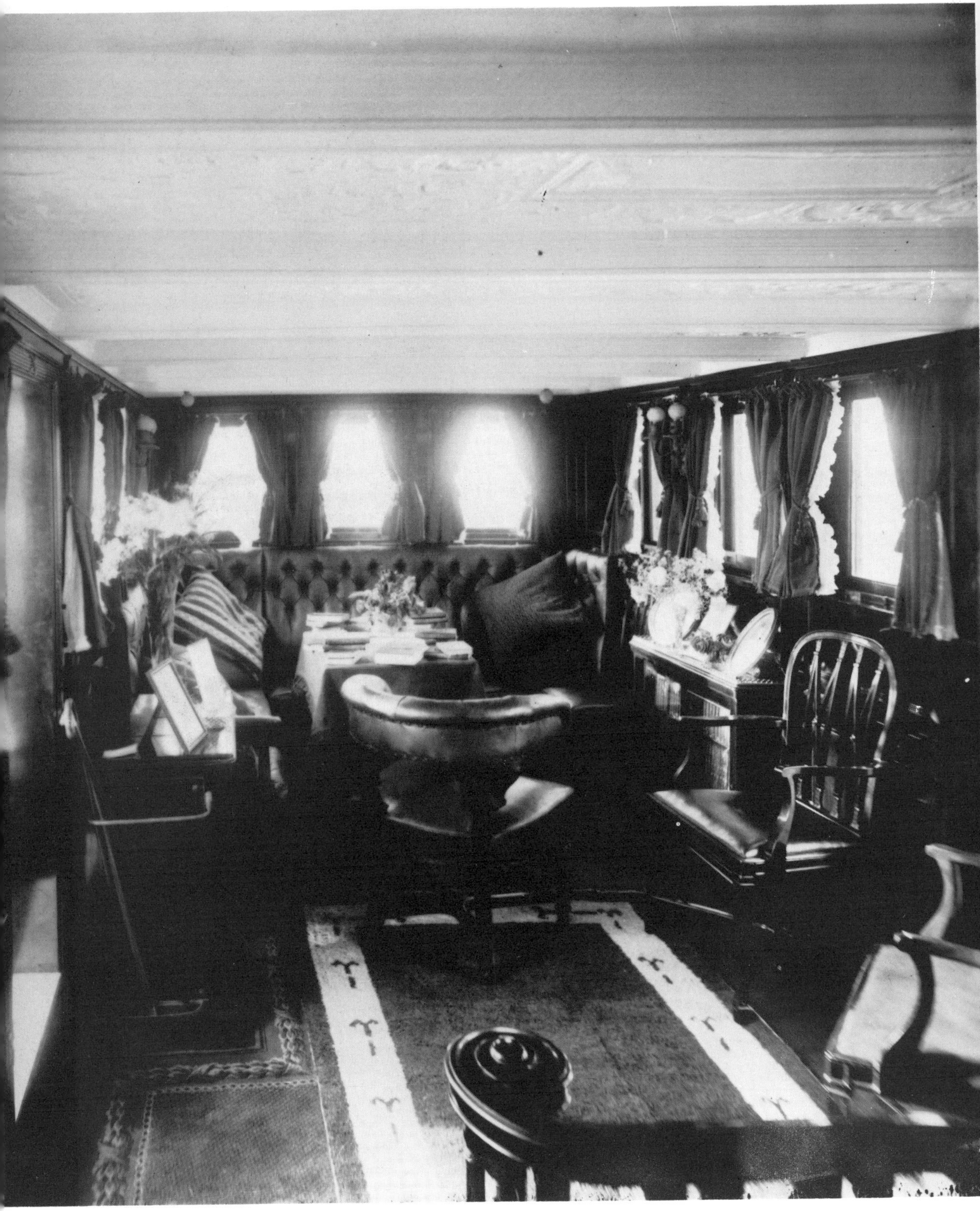

One of the staterooms of the steam yacht *Invincible*, with every comfort provided.

Left A close-up of *Invincible*'s deck saloon, 1910. The steam yacht was built in 1893 in the United States of America.

Miss Squire's bedroom on board the steam yacht *Invincible*. This was typical of a society lady's bedroom on board ship in 1910. Note the beautiful silver dressing-table set and the drawers under the bunk to stow away clothes and blankets, etc.

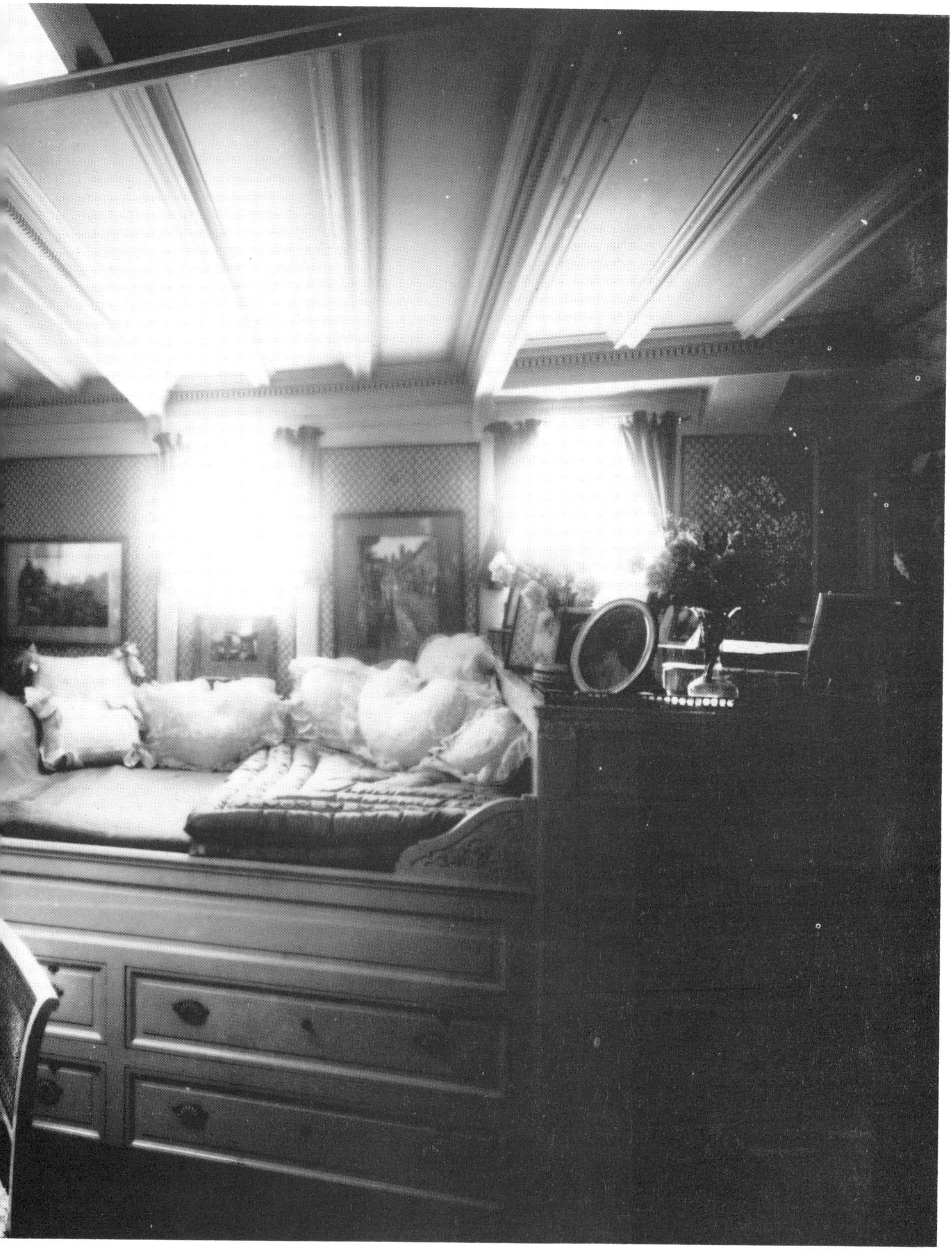

The deck saloon of the steam yacht *Invincible*, 1910. She represented real luxury at sea, with every modern comfort, even electric light. Note the plaster ceiling.

Miss Squire's private room on board the steam yacht *Invincible*, 1910. Note the bath in the left-hand corner.

A splendid steam pinnace built by J. Samuel White for the Russian royal yacht *Pole Star*. The photograph was taken of her sea trials in 1911.

J. Samuel White's yard at Cowes was involved in many foreign contracts to build both large and small vessels. Illustrated here is one of a series of five launches built for the Rio Customs in 1911. Note the framework above the men, to hold a tarpaulin in place during hot weather.

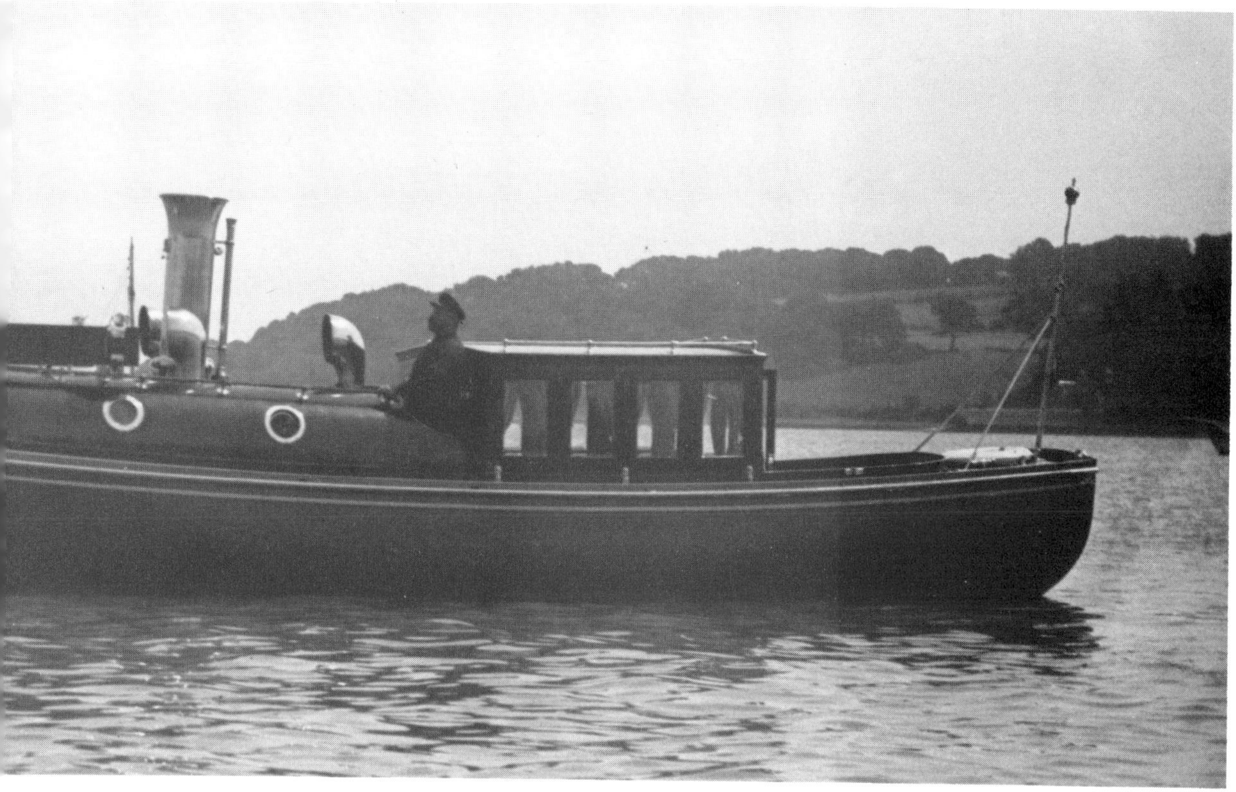

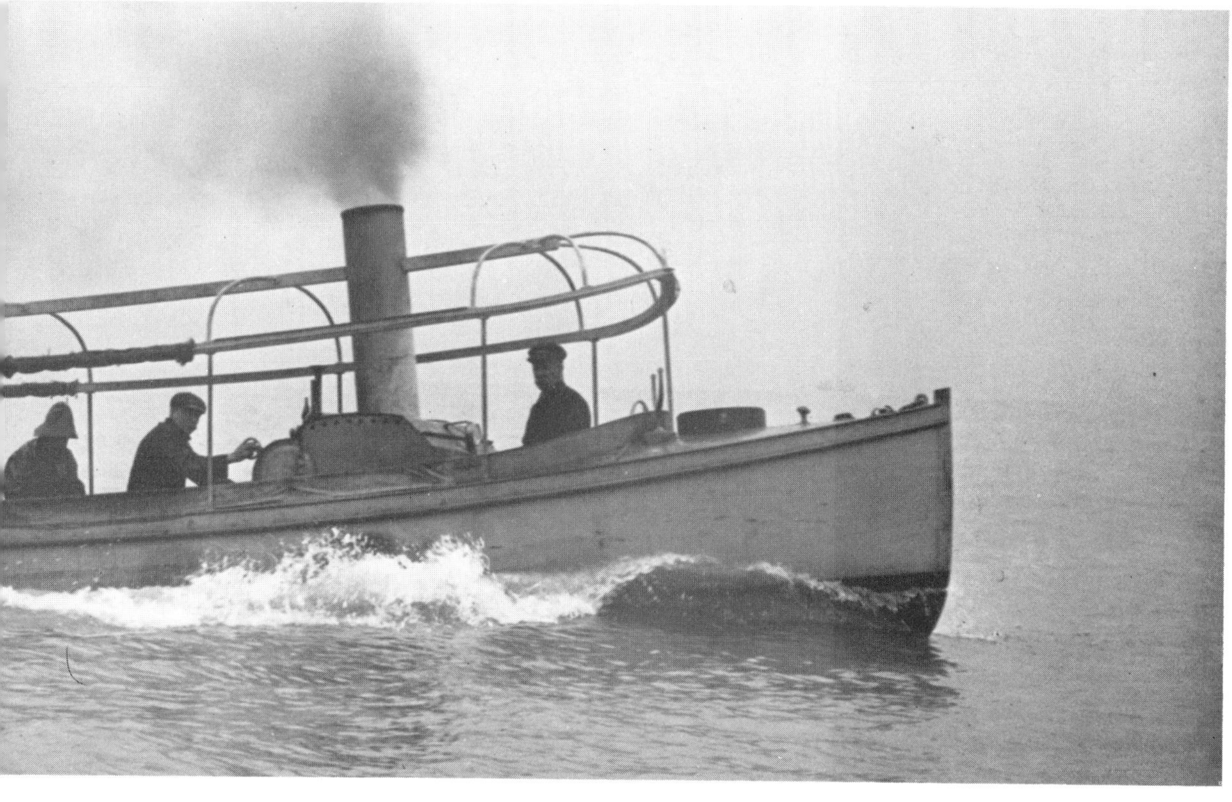

The royal yacht *Victoria & Albert III* underweigh off the Isle of Wight c.1910. She was the last in the line of splendid royal steam yachts to be built before 1914.

Right The Archduke Charles Stephen on board the royal yacht *UI*, with his family, c.1912. *UI* was built by Ramage & Ferguson and had a gross tonnage of 709.

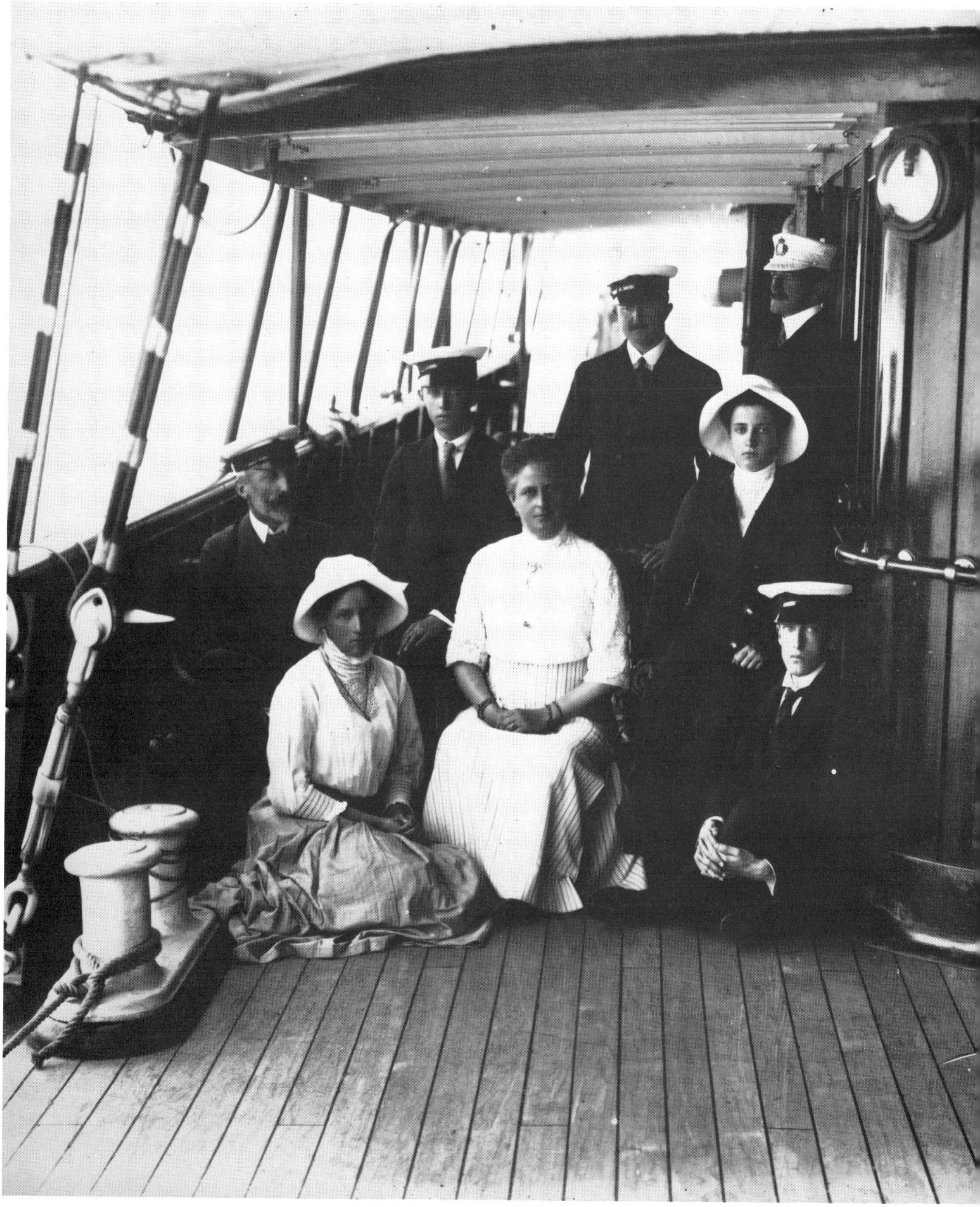

SAPPHIRE 1912

DESIGNER G.L. Watson & Co. BUILT 1912 by John Brown
LOA 285 feet BEAM 35 feet 2 inches DRAUGHT 14 feet GROSS TONS 1207
MATERIAL steel ENGINE two TE 18 × 29 × 32 × 32/27

Sapphire was one of the finest examples of G.L. Watson's work. The yacht was pleasing to look at, and combined first-class comfort with excellent seagoing quality. She was built for the Duke of Bedford, but was to a large extent used by his wife, an ornithologist of repute.

During the First World War, *Sapphire* served as an auxiliary patrol vessel, stationed at Gibraltar, having been armed with one 4-inch gun and one 12-pounder. *Sapphire* again saw service in the Second World War as a submarine tender, under the name of HMS *Breda*.

On 18 February 1944 *Sapphire* was sunk in a collision at Campbelltown Loch, a sad end to a fine yacht.

Sapphire is seen here underweigh on her trials in the River Clyde, 1912.

The steam yacht *Sapphire* in 1912 at John Brown's yard, being fitted out. Behind her is HMS *Southampton*.

The steam yacht *I Wonder* built by J. Samuel White of Cowes, on sea trials in 1912.

NORAH 1913-14

BUILT 1913–14 by Tom Taylor & Sons, Staines, Middlesex
ENGINE compound

Unfortunately, there is little information at hand on *Norah*, but it is thought that she was built for use on the *Tigress*, and *Uphrates* during the Great War.

Norah is seen here at anchor on the Thames, possibly during her trials.

XARIFA II 1927

BUILT 1927 by J.S. White of Cowes   MATERIAL steel   LOA 204 feet 4 inches
BEAM 31 feet 1 inch   DRAUGHT 13 feet   GROSS TONS 731
ENGINE two TE $11\frac{3}{4} \times 18 \times 29/24$ each

Xarifa was one of the last big steam yachts to be built in Britain. Built for R.M. Singer, a well-known yachtsman who had most of his other yachts named the same. R.M. Singer owned *Xarifa* from 1930–47. During this period, *Xarifa* was taken over by the Royal Navy for service in the Second World War. Re-named HMS *Black Bear*, she served in Trinidad. After the war, *Xarifa* was bought for commercial interests and re-named SS *Cymania*.

Taken in 1930, the photograph shows the *Xarifa* with her usual spoon bow, a distinctive feature of the yacht.

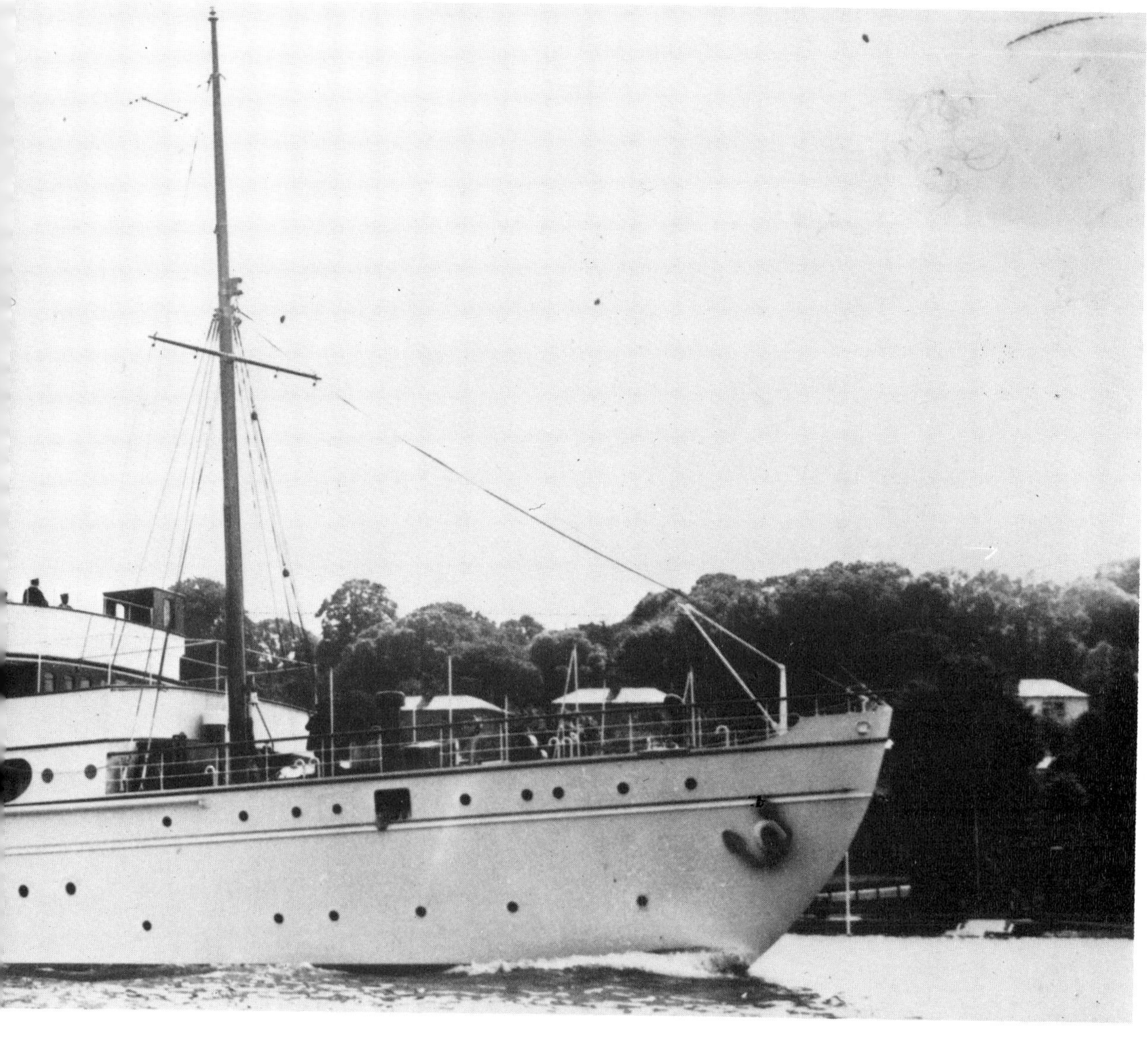

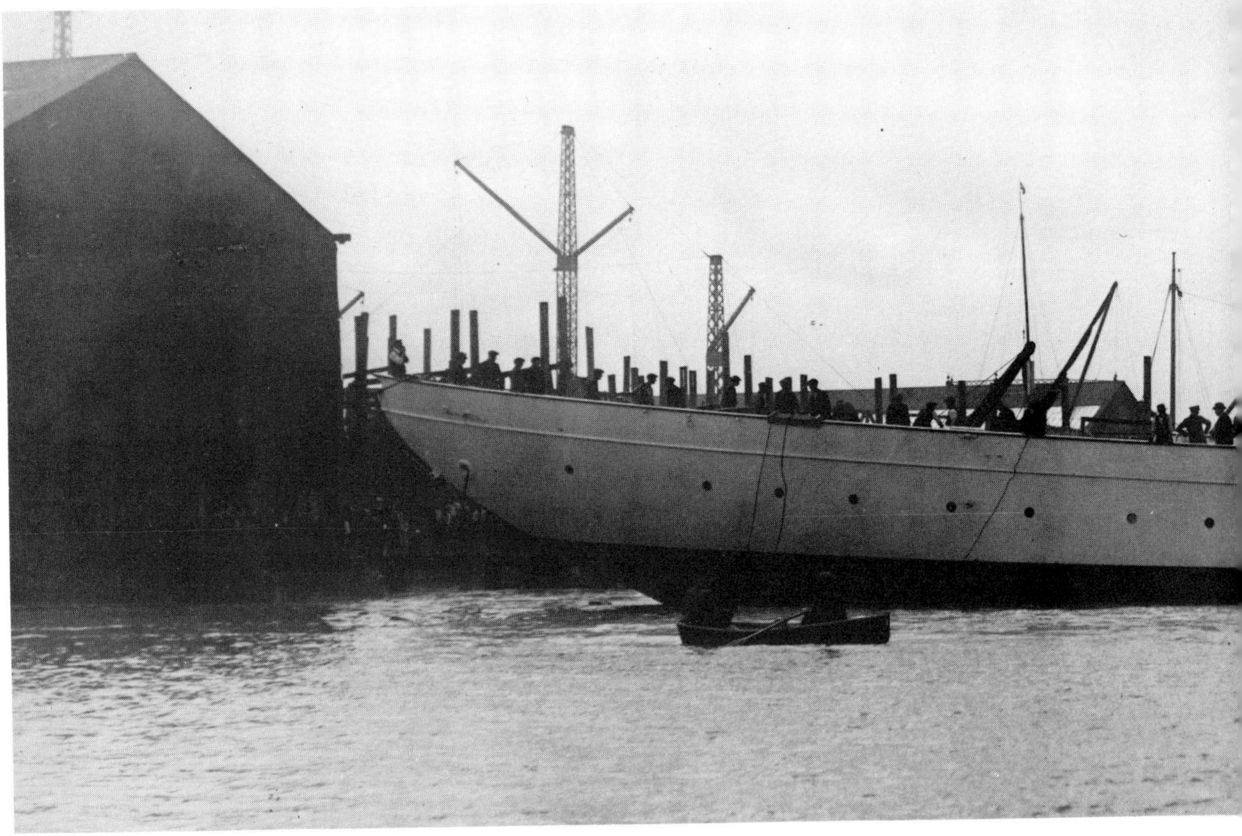

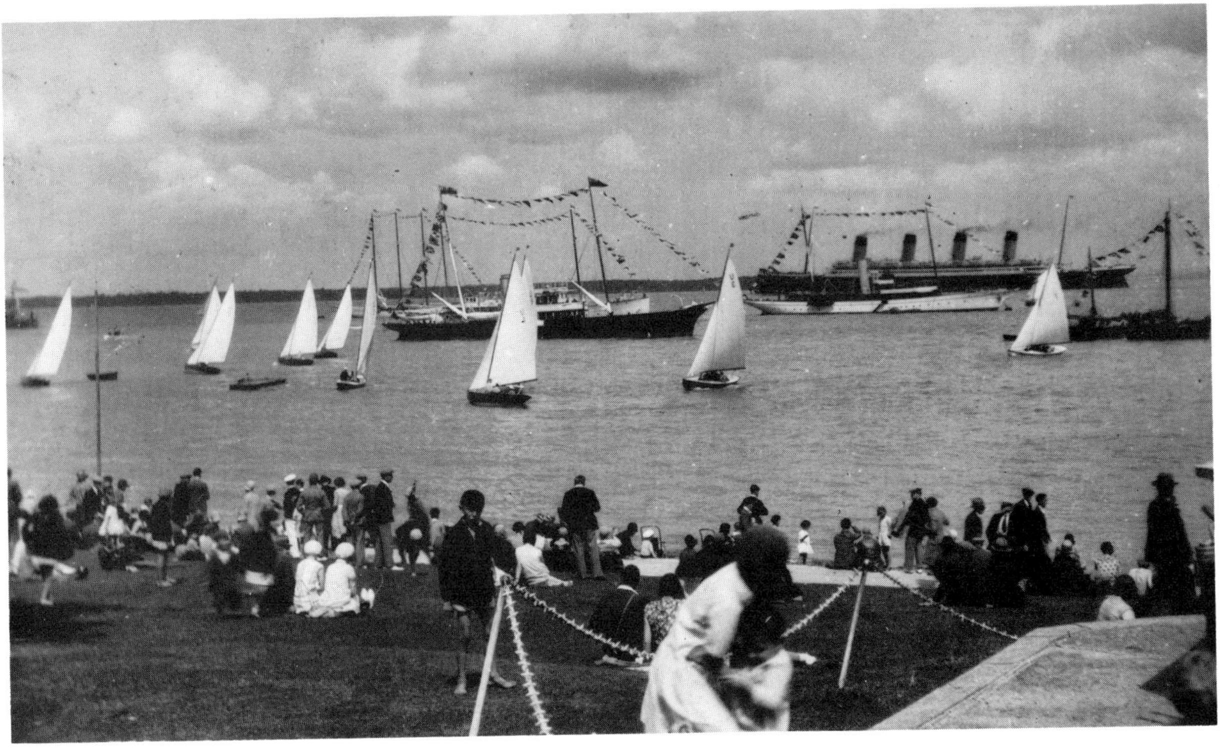

The launching of the steam yacht *Xarifa II* at J.S. White's yard, East Cowes, on 30 June 1927, built for J.M. Singer.

Xarifa II, minus her masts and smoke stack, which had yet to be fitted, was one of the three last big steam yachts to be built in Britain; the other two being *Rover*, built by Alexander Stephens & Sons Ltd of Glasgow, and *Nahlin*, designed by G.L. Watson's firm and built by John Brown Ltd. The *Nahlin* was to earn a place in history as the yacht used by King Edward VIII and Mrs Simpson whilst cruising in the Mediterranean.

Left Cowes Regatta, 1930. The regatta was still very much a spectator sport which only the rich indulged in. It must have been magnificent to see the yachts racing around the stately steam yacht. Beyond, the RMS *Olympic* is passing, en route for New York.

SEA BELLE II 1928

BUILT 1927–8 by J. Samuel White & Co., Cowes
LOA 209 feet   BEAM 33 feet 6 inches   DRAUGHT 16 feet   MATERIAL steel
ENGINE triple expansion 6 Cy(2) $11\frac{3}{4}$, (2) $18 \times$ (2) 29–24   GROSS TONS 300

Steam yacht *Sea Belle II*, 1928, after being constructed and fitted out by J.S. White at the company's yard at East Cowes. *Sea Belle II* was fitted with electric light and wireless. She was built for the Government of the Colony of the Straits Settlements, Singapore. During the Second World War she was hired by the Royal Navy and used as an auxiliary patrol vessel until 1942, when she became a base ship.

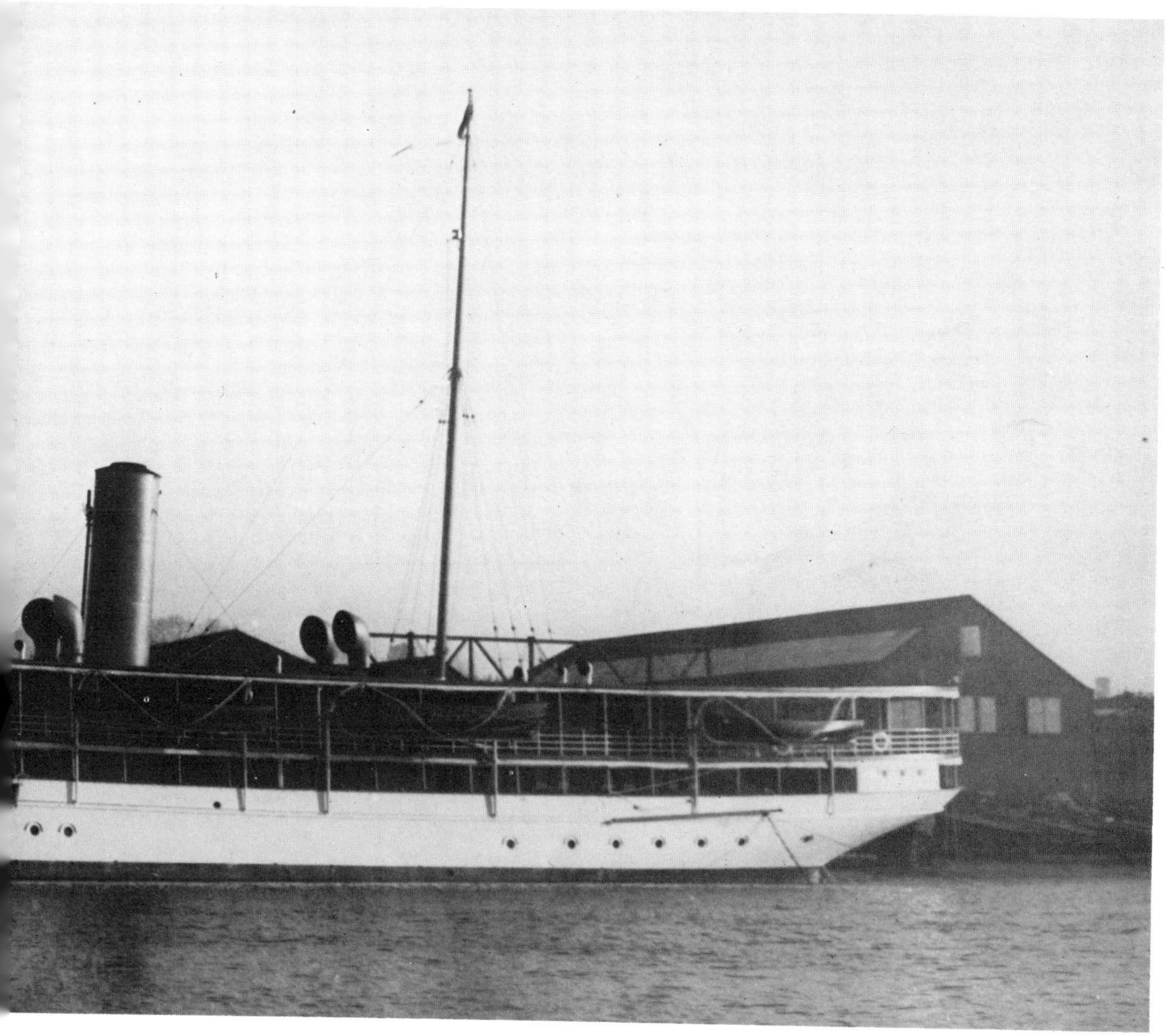

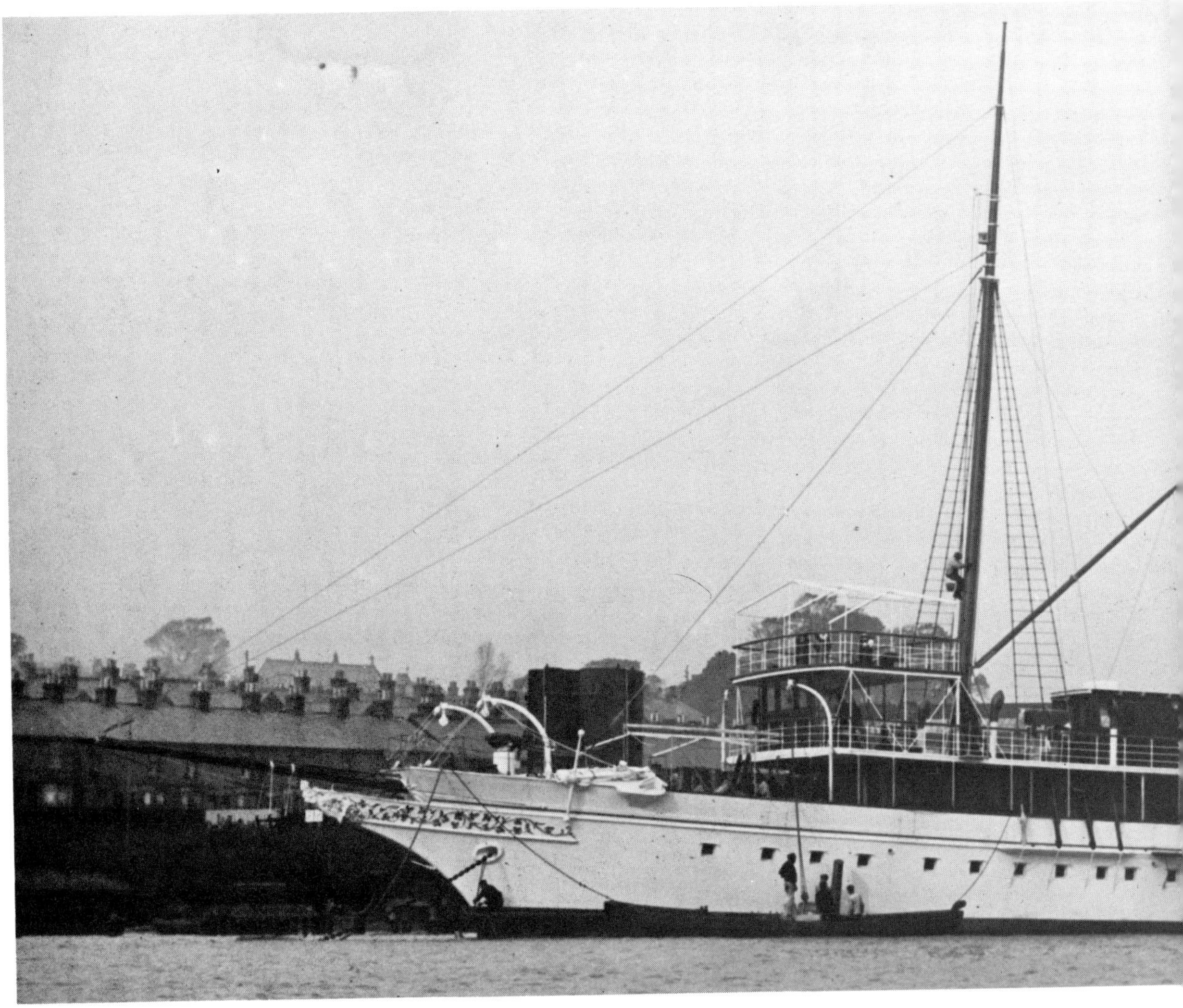

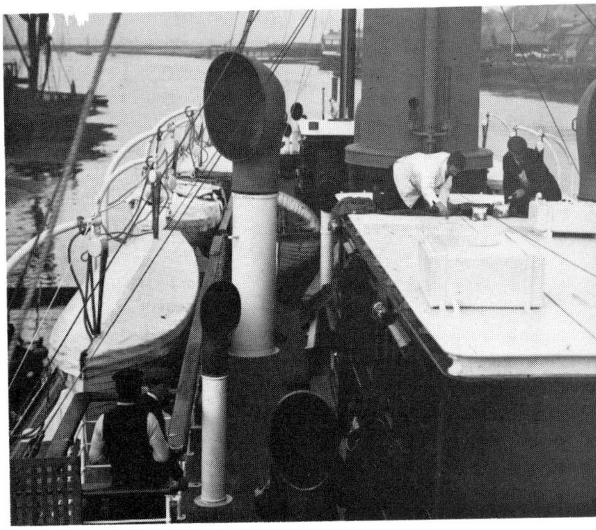

The steam yacht *Ivy* being reconditioned in 1936, by J. Samuel White's works. *Ivy* is berthed at East Cowes, on the river Medina.

Left The *Ivy* being reconditioned at J. Samuel White's yard, Cowes, 1936.

Bibliography

John, Duke of Bedford, *A Silver Plated Spoon*, Cassell, London.
Mrs Brassey, *A Voyage in the Sunbeam*, Lenmans & Co.
Reginald Crabtree, *Royal Yachts of Europe*, David & Charles, Newton Abbot.
Reginald Crabtree, *The Luxury Yacht from Steam to Diesel*, David & Charles, Newton Abbot.
J.J. College, *Ships of the Royal Navy*, David & Charles, Newton Abbot.
J. Grigsby, *Royal Yachts 1604–1953*, Adlard Coles, St Albans.
Erik Hofman, *The Steam Yacht: An Age of Elegance*, Nautical Publishing Co., Lymington.
Hughes, *Famous Yachts*, Methuen & Co., London.
H.T. Lenton and J.J. Colledge, *Warships of World War II*, Ian Allen, Shepperton.
Lloyds Register of Yachts
Royal Yachts, Her Majesty's Stationery Office, Norwich.
Steamboat Catalogue (Reprint No. 3), Gordon Hills & Co.
Heckstall Smith, *Sacred Cowes*, Allen & Wingate, London.
Heckstall Smith, *The 'Britannia' and her Contemporaries*, Methuen & Co., London.
Lady Yarrow, *Alfred Yarrow – His Life and Work*, Edward Arnold & Co., London.

Index

Numbers in italics refer to pages with illustrations

Admiralty steam launch No. 63   *78*
Agatha   86
Alberta   15, *16*, 77
Alexandra   93
Allah Karin   47
HMS *Amazon*   85
Amy   25
Apache   33
Aphrodite   49
Atlantic   33

Beq-Hir   34
Black Bear   110
Blunderbus   23
Breda   106
Britannia   12, 53
Britannia cutter   85, 93
SS *Bute*   48

Carmela   77
Catania   40
Cavalier   22
Christina   12
Chrysoprase   *52*, 77, *82, 83*
Corsair   49
Cuhona   *74, 75*
SS *Cymania*   110

Devonia   41
Dorothea   30
Dorothy   41

Eda   63
Elena   34
Elfen   63
Else   87
Enchantress   15, 77, *79*
Ensa   34
Ethleen   *80, 81*

Fairy   16
Fire King   8
Fire Queen   77
Frivolities   76

Galatea   42
Gelasma   84
Gladiator   30
Gladys   28, 43, 87
Glencairn   95
Greta   41

Hecla   46
Hélène   30
Henriette   29
Hohenzollern   15, 68
Hamburg   33

Invincible   *96, 97, 98, 99, 100, 101*
Irene   77
Irishman tug   85
Isis   15
Ivy   *38, 39, 116, 117*
I Wonder   108

Jade   90
Jason   25

Kanahwa   49

Lady Forrest   *56, 57*
Lady Sophia   *42, 43*
Latona   45
Lensahn   68
Le Tigre   31
Liberty   *94,* 95
Liberty IV   95
Lucille   42
Luciole   43

Maid of Honour   93
Mala   48
Mallard   47
Margarita   22
Marjorie   42
Maxtino   47
Mayflower   48
Medusa   91
Menai   8, 34

Mercedes   31
Mollusc   91
Moonbeam   34
Morven   29
Moschanthi Toqia   40

Nahlin   113
Nere-Naetza   87
Nino Pittaluga   43
Norah   109
Norma   49

RMS *Olympic*   112
Orient   26
Osborne   77
Osborne launch   91

Paladin   35
Pole Star   102
Puffin   44

Rio Customs launch   102
Rover   87
Rover   113

Salamina   26
Sapphire   8, 9, 10, 11, *106, 107*
Scud   21, *44*
Sea Belle II   *114, 115*

Silvia Onorato   43
HMS *Southampton*   107
Star of India   66
Stella   18
HMS *Stormont*   12
Sunbeam   8, 25, 33
Sunrise   24, *88, 89*
Sylvana   92

Tchantches   35

UI   105

Valfreyia   66
Valhalla   8, *32, 33*
Victoria & Albert II   *14*, 16, 20, *34, 71*
Victoria & Albert III   12, *34, 52, 53, 71*, 77, 93, *104*

Wildfire   31
HMS *Winchelsea*   85
Windsor Castle   14

Xarifa   *36, 37*
Xarifa II   *110, 111, 112, 113*

Yves de Kerguelen   24

Zaida   *58, 59, 60, 61, 62*
Zaranda   41